Sustainable Finance

Series Editors

Karen Wendt, CEO. Eccos Impact GmbH
President of SwissFinTechLadies, President Sustainable-Finance
Cham, Switzerland

Margarethe Rammerstorfer, Professor for Energy Finance and Investments
Institute for Finance, Banking and Insurance WU Vienna
Vienna, Austria

Sustainable Finance is a concise and authoritative reference series linking research and practice. It provides reliable concepts and research findings in the ever growing field of sustainable investing and finance, SDG economics and Leadership with the declared commitment to present the theories, methods, tools and investment approaches that can fulfil the United Nations Sustainable Development Goals and the Paris Agreement COP 21/22 alongside with de-risking assets and creating triple purpose solutions that ensure the parity of profit, people and planet through choice architecture passion and performance. The series addresses market failure, systemic risk and reinvents portfolio theory, portfolio engineering as well as behavioural finance, financial mediation, product innovation, shared values, community building, business strategy and innovation, exponential tech and creation of social capital. Sustainable Finance and SDG Economics series helps to understand keynotes on international guidelines, guiding accounting and accountability principles, prototyping new developments in triple bottom line investing, cost benefit analysis, integrated financial first plus impact first concepts and impact measurement. Going beyond adjacent fields (like accounting, marketing, strategy, risk management) it integrates the concept of psychology, innovation, exponential tech, choice architecture, alternative economics, blue economy shared values, professions of the future, leadership, human and community development, team culture, impact, quantitative and qualitative measurement, Harvard Negotiation, mediation and complementary currency design using exponential tech and ledger technology. Books in the series contain latest findings from research, concepts for implementation, as well as best practices and case studies for the finance industry. This book series is indexed in SCOPUS

Andrea Quintiliani

ESG and Value Creation from a Corporate Finance Perspective

Integrating Sustainability into Financial Strategy

Springer

Andrea Quintiliani
University of Chieti-Pescara
Pescara, Italy

ISSN 2522-8285 ISSN 2522-8293 (electronic)
Sustainable Finance
ISBN 978-3-032-03146-4 ISBN 978-3-032-03147-1 (eBook)
https://doi.org/10.1007/978-3-032-03147-1

This Springer imprint is published by the registered company Springer Nature Switzerland AG
The registered company address is: Gewerbestrasse 11, 6330 Cham, Switzerland

To my mother, Nicoletta. Although you are no longer by my side, your memory and spiritual presence continue to guide and sustain me at every step of my journey.

Contents

About the Author

Andrea Quintiliani is Assistant Professor of Corporate Finance at the University G. d'Annunzio of Chieti-Pescara, Department of Management and Business Administration, where he teaches Corporate Finance in both the bachelor's and master's degree programs. He has held several academic positions, including Full Professor and Researcher in Financial Intermediation and Corporate Finance. His research focuses on corporate finance, the bank–firm relationship, internal credit rating systems, business valuation, and financial sustainability. In addition to his academic career, he has professional experience as a financial and business consultant, specializing in economic and financial analysis, business valuation, and corporate financial management. He has a particular focus on small and medium-sized enterprises (SMEs) and their relationships with financial institutions. The author of numerous scientific publications on corporate finance, he has actively contributed to national and international research projects and serves on the editorial boards of leading academic journals in finance and economics.

Chapter 1
Introduction

The acronym ESG (Environmental, Social, Governance) undoubtedly evokes one of the most current and debated topics in the investment field.

Over the past 3 years, countless individuals have turned to Google to search for the meaning of "ESG investments". The level of attention is extraordinarily high.

The impetus generated by the Covid-19 pandemic regarding social issues, together with growing environmental concerns justified by the phenomenon of global warming, has compelled business leaders to engage seriously with environmental, social, and corporate governance matters; the latter was already under scrutiny by the institutional investment community even before the outbreak of the health crisis.

But what is it? Is ESG investing a fad or a long-term trend that will dominate the capital market for the foreseeable future?

As highlighted in a previous scientific contribution (Quintiliani, 2021), ESG is undoubtedly a trend that has been underway for some time. However, it has undergone a strong acceleration with the health emergency.

The succession of different epidemic waves and confinement measures has pushed companies to reduce their carbon footprints.

The progressive reduction in commuting, the spread of smart working, the emergence of a more careful awareness of the severity of an epidemic event, the now sedimented fear of an invisible enemy constantly lurking, are all factors that, post-Covid, are encouraging companies to calibrate and manage their impact in environmental, social and governance terms.

As can be seen from the survey conducted by Welsh (2018) on a significant panel of American investors, already in 2018 activist shareholders showed a certain sensitivity in supporting ESG objectives and initiatives: climate change (19%), environmental sustainability (13%), other environmental initiatives (7%) and policies (19%). Based on the empirical evidence obtained by Welsh (2018), it is clear that 39% of the resolutions presented by shareholders were on environmental matters.

A. Quintiliani, *ESG and Value Creation from a Corporate Finance Perspective*, Sustainable Finance, https://doi.org/10.1007/978-3-032-03147-1_1

It is worth noting that in commenting on these numbers, we do not want to express a value judgment; the objective is to show that environmental issues are gaining the interest of a plurality of investors and, most likely, will significantly influence the world of economics and finance.

The world of finance is inspiring its choices also thanks to the adoption and use of ESG rankings (Bradley, 2021). In this regard, the study conducted by Bellavite Pellegrini et al. (2020) highlights the significant impact of ESG scores on the cost of capital.

Recent academic literature, concerning the study by Dallocchio et al. (2025), emphasises the importance of the capital structure of companies in the context of climate transition. Proactive companies in managing climate risks[1] show a greater ability to quickly adjust their optimal financial leverage, benefiting from better access to credit and a lower cost of capital (Muchenje et al., 2025).

It is worth remembering how companies often benefit from activities that do not give rise to financial outflows but that determine external costs. These costs, which are not accounted for but transferred to society in general, are called externalities. Therefore, business activity can be a source of numerous negative externalities with a high environmental and social impact.

The health emergency must not be a conditioning element but a stimulus for all companies, so that they tend to estimate negative externalities, both environmental and social.

Following the outbreak of the pandemic, rating agencies communicated to companies that the presence (or absence) of sustainability plans and objectives could be reflected in the prices of their bonds and their WACC—Weighted Average Cost of Capital (Schenker, 2020). In other words, it could impact profitability, enterprise value, creditworthiness, and share price.

In the future, companies will increasingly need to demonstrate the achievement of ESG goals. Otherwise, they may find themselves at the mercy of activist shareholders and, therefore, be negatively affected in terms of reduced profitability and share price.

Given the foregoing, it is important to clarify that the literature on ESG issues—hundreds of articles produced in the last decade alone—can create disorientation in corporate finance scholars, who see the multiple aspects of the ESG world addressed in an inconsistent, confused and, at times, promiscuous way (Gillan et al., 2021).

In corporate finance studies, the ESG phenomenon should be seen as a two-sided coin:

---

[1] Climate risks are a specific subset of environmental risks (the E in ESG). They relate exclusively to the impacts of climate change, and are typically divided into:

- *Physical risks*: These arise directly from the impacts of climate change, such as extreme weather events (floods, droughts, hurricanes), rising sea levels, and increasing temperatures.
- *Transition risks*: These are associated with the shift towards a low-carbon economy, including regulatory changes, technological innovations, changes in consumer preferences and resulting financial effects.

- On one side, we find the topic of corporate investments based on ESG factors. Companies invest in ESG with the aim of improving their appeal on the credit and equity markets. Furthermore, investments in ESG factors mitigate the risks perceived by investors and, therefore, reduce the cost of capital. Finally, these investments improve economic and financial performance. Investments in ESG factors positively impact the Return On Assets (ROA), the Return On Equity (ROE), Profit Margins, and Cash Flows. Effective management of ESG risk, in particular that attributable to climate issues, therefore emerges as a crucial factor for corporate financial sustainability. Companies that integrate these risks into their strategies show not only a lower exposure to regulatory risks but also a significant competitive advantage in the capital market.
- The other side of the ESG coin is represented by investors (funds) who put sustainability at the heart of their investments and who select portfolios based on the characteristics of the ESG factors integrated into them.

This book aims to delve deeper into the first point in the list, which sees the company investing in ESG factors with the primary intent of increasing its value. Value, the one just stated, which is triggered by ESG investments and attributable to three main areas: market appeal, reduction of risk and cost of capital, and improvement of economic-financial performance.

Regarding market appeal, it should be made clear that the integration of ESG criteria into corporate strategies has become an essential lever for attracting capital, investors, and clients, as more and more stakeholders favour companies committed to sustainability and social responsibility. Institutional investors, in particular, value companies with good ESG scores, which are seen as less risky and better able to address environmental and social challenges. ESG also offers a competitive advantage on the stock market: companies that meet high sustainability standards often register higher market valuations and enjoy a good reputation. This green and responsible positioning allows companies to stand out in highly competitive sectors, and also gains them more visibility in major sustainability indices such as the Dow Jones Sustainability Index or the MSCI ESG Index, which further increases equity appeal and facilitates access to capital.

On the risk reduction side, ESG investments help reduce corporate risks on multiple fronts, mitigating reputational, operational, environmental, and regulatory risks, as well as impacting the cost of capital. Companies with high ESG scores tend to experience lower volatility in financial performance, as they demonstrate more attentive management of long-term risks, including climate change, human rights, and governance risks. This prudent and far-sighted management attracts financiers, as they perceive a lower probability of sudden adverse events that could impact the value of the firm. As can be seen from the study by Dallocchio et al. (2025), ESG investments therefore acquire strategic relevance in mitigating financial risk and improving economic and financial performance. The research shows how companies that are more attentive to environmental factors, particularly climate risk, can outperform through more effective management of operating costs and greater resilience to external shocks.

In terms of cost of capital, studies indicate that companies with sound ESG policies often access financing with more favourable interest rates (lower cost of debt). This occurs because financial partners perceive the company as a less risky and more stable investment in the long term. Furthermore, the cost of risk capital (equity) is mitigated, as equity investors tend to require lower returns on ESG investments, seeing them as less exposed to structural risks. This reduces the WACC, increasing the company's ability to undertake further investments. In particular, according to Dallocchio et al. (2025), the cost of capital can vary significantly depending on the ability of companies to address and anticipate regulatory and climate changes. Companies that perform well in terms of ESG tend to enjoy more favourable financing conditions, which reduces the perceived risk to investors. This is supported by a study by Bannier et al. (2022), which found that ESG transparency reduces information asymmetries and risk premiums, positively influencing the cost of both debt and equity. Recent empirical evidence (Quintiliani, 2024) corroborates this, demonstrating that firms with higher ESG scores benefit from superior market valuations and reduced financing constraints, particularly within sectors subject to environmental regulation.

In terms of improving economic and financial performance, it is clear that ESG investments not only reduce costs but also improve the economic and financial performance of companies. Here are some of the main impacts:

- *ROA*. ESG factors promote more efficient management of resources, helping companies to minimise waste and operating costs, with positive effects on returns.
- *ROE*. Improving efficiency and enhancing corporate reputation can increase stakeholder loyalty and support, positively reflecting on ROE.
- *Profit Margins and Cash Flows*. Sustainable and socially responsible management tends to improve customer loyalty and employee engagement, reducing turnover and acquisition costs and enhancing profit margins.

Finally, the analysis by Dallocchio et al. (2025) highlights that a company's ability to respond, particularly to climate risks, has a positive impact on its economic and financial performance, facilitating a more rapid adjustment of its capital structure and improving its overall evaluation by stakeholders.

In summary, the adoption of ESG criteria proves to be a strategic approach not only for meeting sustainability expectations but also for achieving tangible financial and operational benefits. With the increase in regulations and expectations from customers and investors, companies that do not integrate ESG criteria risk finding themselves at a competitive and financial disadvantage in the medium to long term.

In light of the above, the integrated management of ESG risks is not only an ethical consideration but also a functional strategy for achieving financial resilience and ensuring long-term competitive survival. In line with this, Broadstock et al. (2021) found that Chinese companies with higher ESG scores showed greater resilience in financial markets during the pandemic crisis, confirming the protective role of sustainable practices in contexts of systemic stress. Quintiliani's (2024) findings echo these considerations, indicating that firms with robust ESG practices exhibited greater financial resilience during periods of macroeconomic stress. This confirms the strategic function of ESG as a buffer against volatility.

This volume is divided into eight chapters, each of which addresses in a systematic and progressive way the interrelationships between ESG criteria and corporate finance, to provide a scientifically rigorous, up-to-date, and operationally relevant treatment.

Chapter 2—*ESG principles and conceptual foundations*—introduces the conceptual foundations of the topic, clarifying the meaning of ESG criteria, their origins, and the distinction from similar approaches such as Corporate Social Responsibility (CSR), Corporate Sustainability (CS), and Impact Investing (II). The evaluation system of ESG indicators is also analysed, concerning the growing importance of double materiality and current European legislation.

Chapter 3—*ESG metrics, indices and financial instruments*—delves into the methods of construction, detection, and use of ESG indicators, focusing on ESG indices (equity and bond), their financial implications, and innovative sustainable finance instruments, such as green bonds, sustainability-linked bonds, and loans. Particular attention is paid to the problems of heterogeneity and comparability between ESG ratings and the consequent difficulty of benchmarking.

Chapter 4—*ESG factors and firm financial performance*—represents the analytical heart of the volume. Here we explore in depth the connections between ESG performance and the main economic-financial dimensions of the company: financial performance, creditworthiness, cost of capital (debt and equity), firm valuation, and capital structure. The intent is to understand whether and to what extent the integration of ESG factors generates value for the company and for investors.

Chapter 5—C*apital structure, risk management and ESG strategies*—extends the analysis to the strategic aspects of financial management. Capital structure decisions are analysed from a sustainable perspective, the integration between ESG risks and financial risks, corporate resilience as an outcome of ESG strategies, and the role of ESG rating in determining the cost of capital. The chapter also provides a critical reading of the long-term effects of ESG integration on the competitive sustainability of the company.

Chapter 6—*Literature review (years: from 2020 to 2025—Journal: indexed Scopus & Wos)*—offers a structured and thematically organised review of international academic literature, published in scientific journals indexed by Scopus and Web of Science. The analysis is divided into five areas: (i) ESG and financial performance (15 papers); (ii) ESG and credit rating (2 papers); (iii) ESG and cost of capital (5 papers); (iv) ESG and firm value (6 papers); (v) ESG and capital structure (2 papers). The aim is to offer a robust and up-to-date reference framework for the theoretical and applicative reflections developed in the previous chapters.

Chapter 7—*Empirical studies and applications*—presents three original empirical contributions, focused on the link between ESG and corporate value, ESG and financial performance, as well as on the impact of FinTech technologies on SMEs' ESG strategies, with particular reference to the Italian and British context. The results of these investigations provide concrete and measurable evidence for the conceptual hypotheses previously discussed.

The volume concludes with a final section (Chap. 8) entitled *Final reflections on ESG and Corporate Finance*, which summarises the main findings, offers critical insights, and provides directions for future research and the evolution of managerial practices in the field of sustainable finance.

In continuity with the theoretical and applicative framework outlined in the chapters, the volume intends to propose an interpretation of sustainability not as an external constraint to the economic-financial logic, but as a transformative and value-generating principle. The centrality attributed to double materiality, to the interaction between ESG risk and financial risk, as well as to the role of ESG factors in determining the cost of capital, demonstrates the desire to overcome a normative or reputational reading of sustainability, in favour of an integrated and performance-oriented vision. The objective pursued is not only to document the growing impact of sustainability on corporate choices, but also to offer an analytical framework capable of guiding scholars, managers, and policy makers in understanding the new determinants of corporate value in a rapidly evolving regulatory, financial, and competitive context. The integration of ESG factors in corporate strategies, financial decisions, and governance models emerges as a critical lever to strengthen the resilience of the company, increase its attractiveness on the markets, and ensure competitive sustainability in the long term. The hope is that this study may represent a useful tool not only to orient oneself in the debate, but also to act consciously within an economic context undergoing profound transformation.

Chapter 2
ESG Principles and Conceptual Foundations

2.1 The ESG Paradigm: Meaning, Evolution, and Differentiation from CSR and Sustainability

Commonly defined as the analysis of a company's environmental, social, and governance practices, ESG first caught the attention of the financial world following a 2005 United Nations Global Compact report that argued that incorporating ESG criteria into capital markets would enable companies to do well by doing good (Bradley, 2021). Since then, the importance of ESG issues has skyrocketed.

In confirmation of the strong interest in ESG issues, it is useful to remember how the international network of investors, certified as per the Principles for Responsible Investment (PRI), today counts more than 1200 signatories among institutional investors, asset management companies, and service providers.

Launched by the United Nations (UN) in 2006, the "Principles for Responsible Investment" aim to promote the diffusion of sustainable and responsible investment among institutional investors.

Investors' heightened sensitivity to the issue of environmental impact and their greater awareness of the critical role of ESG performance in safeguarding a company's success mean that ESG can no longer be considered a niche investment.

ESG issues may represent both opportunities and risks for companies and, therefore, determine their long-term prospects.

In recent years, the term ESG has become synonymous with sustainable investment or, better yet, socially responsible investment. However, ESG, rather than being a stand-alone investment strategy, should be better interpreted and conceptualised as part of a risk management system that is functional to the analysis of firm value. Integrating risks related to ESG factors allows the company to better understand the context in which it operates, to be more aware of the risks taken, and, ultimately, to strengthen its ability to respond to stakeholders' needs.

A. Quintiliani, *ESG and Value Creation from a Corporate Finance Perspective*, Sustainable Finance, https://doi.org/10.1007/978-3-032-03147-1_2

For greater clarity, the acronym ESG indicates the criteria suitable for measuring the sustainability and social impact of an investment in a company. These criteria prove to be strategic because, in perspective, they allow company management to better align with financial performance objectives.

It is important to underline that ESG should not be confused with Impact Investing (II). This refers to the type of investment to which the investor is directed, while ESG criteria are part of an evaluation process that borrows non-financial data to identify the tangible risks and growth opportunities of the company. Furthermore, with II, the investor intentionally aims to obtain economic returns from an investment suitable for generating a positive, and therefore measurable, social and environmental impact. Instead, unlike Impact Investing, ESG is a means to an end, as it is suitable for identifying those non-financial risks that, ultimately, can be reflected, in terms of tangible impact, on the value of the company.

Furthermore, ESG is often mistakenly equated with two factually different phenomena: Corporate Sustainability (CS) and Corporate Social Responsibility (CSR). Although there are some overlaps, these terms cannot be considered interchangeable:

Although there is some overlap, these terms are not interchangeable:

- *CS* is generally used to describe a company's ability to seize opportunities and manage risks arising from economic, environmental and social developments, to create long-term value for all its stakeholders. For many others, corporate sustainability simply means doing good and does not require pre-established conditions.
- *CSR* refers to the integration by companies of the social and environmental concerns of key stakeholders into their operations and business activities of interest. Such actions are undertaken voluntarily by companies. In comparison, ESG evaluates a company's ESG practices in conjunction with more traditional financial metrics (Martielli et al., 2025).

Finally, ESG is further associated with ethical investments. However, ESG criteria configure a preparatory approach to investment which explicitly takes into account environmental, social and corporate governance factors.

In other words, ESG provides an investigation framework suitable for capturing the individual factors (pillars) E, S and G, the most relevant risks and opportunities that companies will have to face. This leads investors to select the stocks that boast the best ESG scores (high-ranked or best-in-class) and, conversely, to reject those stocks that, for example, record an environmental score not in line with their expectations.

It is important to point out that in recent years, the concept of dual materiality, introduced by the European Commission (Adams et al., 2021), has gained centrality. It distinguishes between financial materiality (ESG impacts on economic performance) and impact materiality (the company's impacts on the environment and society). This distinction is crucial for integrated reporting and for alignment with the Sustainability Accounting Standards (SASB) and the European Sustainability Reporting Standards (ESRS). For greater clarity, dual materiality broadens the scope of traditional analysis, overcoming the exclusively shareholder-centred approach. In this sense, ESG becomes an accountability tool towards all stakeholders and not just towards investors. It is now the basis of the Corporate Sustainability Reporting Directive (CSRD), which requires European companies to report ESG risks and impacts according to an integrated logic.

In the contemporary debate in corporate finance, the notion of financial materiality is establishing itself as a central analytical category, especially concerning the integration of ESG factors in the economic and financial assessments of companies. Unlike traditionally understood materiality (impact materiality), which refers to the relevance of information for investors' decisions, financial materiality in the ESG context requires a more detailed reflection on the link between ESG risks and economic performance, in a medium-long-term perspective. One of the most influential contributions to the topic is the study by Khan et al. (2016), which introduced a fundamental distinction between material and immaterial ESG practices based on the sectoral relevance of the issues addressed. The authors empirically demonstrate that companies focused on ESG aspects considered materially relevant for their sector achieve superior financial performance compared to those focused on generic or less relevant ESG pillars. This study has started a line of research that tends to evaluate sustainability not as a homogeneous set of virtuous practices, but as a selective and targeted set of factors potentially generating economic value. Starting from this distinction, academic contributions that delve into the link between financial materiality and ESG disclosure have multiplied. A significant strand focuses on the information effects deriving from ESG reporting based on materially relevant issues, demonstrating that it reduces uncertainty perceived by investors and contributes to a lower dispersion of earnings forecasts (Christensen et al., 2022). The results of this research suggest that materiality should not be considered exclusively as an accounting category, but as a strategic reporting tool capable of influencing the perception of risk and, consequently, the cost of capital.

A study by Huang (2019) shows how the alignment of ESG disclosures with financial materiality priorities can improve the effectiveness of the dialogue with institutional investors and, in particular, promote active engagement strategies based on the relevance of social and environmental impacts on the value of the company. This orientation is also reflected in the evaluation models developed in the academic field to measure the economic significance of ESG metrics in specific sector contexts.

The role of SASB standards—now incorporated into the ISSB architecture—has been the subject of numerous empirical studies that have tested their ability to discriminate between truly significant ESG information and accessory information. For example, Leuz and Wysocki (2016) highlight that adherence to the materiality criteria defined by the SASB standards is associated with greater market reactivity on the occasion of ESG disclosures, confirming the perceived financial relevance of such information. The effect is particularly pronounced in regulated or capital-intensive sectors, where the reputational and operational risk linked to ESG factors is more directly monetisable.

Some studies focus on the role that corporate governance plays in identifying and managing ESG materiality. In particular, Alareeni and Hamdan (2020) have shown that the presence of internal ESG committees and directors with specific expertise in the field of sustainability significantly increases the probability that corporate disclosure is actually oriented towards financial materiality. This suggests that the integration of ESG factors in decision-making processes requires a solid and conscious governance, capable of discerning between greenwashing strategies and approaches truly oriented towards the creation of sustainable value (Hu et al., 2025).

The international comparative horizon has allowed us to extend these reflections to different economic and cultural systems. The comparative study by Ioannou and Serafeim (2017) shows that the effectiveness of ESG disclosure based on financial materiality varies significantly depending on the institutional context, the strength of the regulatory framework and the degree of maturity of the capital market. In highly regulated environments, ESG reporting practices oriented towards materiality are better understood by investors and more effectively rewarded in terms of market valuation.

Recently, literature has also focused on the predictive capacity of ESG materiality in the presence of systemic shocks. The study by Walker et al. (2024) indicates that companies with ESG disclosure focused on materially relevant topics show greater resilience in crisis contexts, such as pandemics or climate disasters. This effect, interpreted as a form of strategic insurance, reinforces the position according to which financial materiality is a key dimension to understand the role of ESG investments in corporate risk management.

Overall, it clearly emerges that financial materiality, far from being a purely formal criterion, represents a critical variable in modelling the relationships between sustainability and financial performance. Its application allows us to overcome generically normative approaches to orient ESG analysis towards a logic of concrete and measurable economic value, consistent with the theoretical and methodological approach that runs through this volume.

In light of the arguments presented, it seems useful to point out that the intensification of the European regulatory framework on sustainability, culminating in the CSRD and the ESRS, cannot be fully understood without considering the driving role played by the pandemic crisis. The experience of Covid-19 has acted as a catalyst for a profound redefinition of the concept of systemic risk and has highlighted the structural interdependence between environmental, social, and economic-financial dimensions. As emphasised by Quintiliani (2021): "*The pandemic has clearly shown the interdependence between public health, business continuity, economic balance and collective responsibility. This has highlighted the unavoidable need for companies to integrate not only economic objectives, but also environmental and social ones into their strategies*". Companies have found themselves faced with the urgency of reviewing their governance model, addressing new demands for transparency, traceability and accountability to all relevant stakeholders. In particular, Quintiliani's study (2021) highlights how the pandemic has shifted the attention of public opinion, regulators and markets to a growing demand for substantial accountability, contributing significantly to the affirmation of a more advanced and structured European regulatory standard. The CSRD, from this perspective, does not represent a simple technical update of non-financial reporting, but the outcome of a paradigm shift triggered by a global crisis that has imposed a profound rethinking of the function of the company in society.

2.2 Evaluation Systems and Regulatory Frameworks for ESG Indicators

Unlike traditional financial ratios, there is no single ESG indicator evaluation system shared by practice and literature. The most recent literature highlights a wide heterogeneity among ESG ratings provided by different providers, with negative effects on the benchmarking capacity and the quality of investment decisions (Berg et al., 2022). In other words, it is not yet clear which rating systems are able to quantify, with precision, the levels of ESG performance.

For clarity, the ESG rating is a report that certifies the solidity of the company from the point of view of environmental, social and governance aspects.[1] In assigning the rating, another question, still open, is the opportunity to integrate all three ESG factors or only some. In fact, for certain stocks, only some of the ESG factors could prove significant.

For example, the environmental risks associated with the activity of a bank will be less significant than those that a company operating in the steel sector will have to face. On the other hand, the hypothesis that sees the financial intermediary more exposed to risks deriving from factors attributable to governance issues seems plausible.

A further aspect, not to be overlooked, is the rebalancing of equity portfolios through the integration of ESG criteria. This integration, functional to the maximisation of value and the minimisation of risk, is consistent with the fiduciary duty, proper to managers and institutional investors, to take into consideration all relevant information, material risks and opportunities that characterise potential investments. Concerning this aspect, it is worth remembering that the approach to ESG integration should not be interpreted as automatically excluding (negative screening) or including (positive screening). The integration of ESG criteria, unlike fundamental analysis, is not constrained by the need for portfolio diversification and can include stocks that, despite registering low ESG ratings, are considered, in perspective, to be performing as their issuing companies demonstrate a commitment to reconsidering their objectives and responsibilities in terms of ESG.

[1] An ESG rating is a score that evaluates a company's commitment to environmental, social and governance issues, and how proactively it manages the ESG issues that are most relevant to its business.

It follows that all large institutional investors are directed to use ESG factors. In particular, the signatories of the PRI (Principles for Responsible Investment) have made efforts to promote the incorporation of environmental, social and corporate governance factors into investment decision-making processes.

For example, BlackRock, the world's largest investment manager, headquartered in New York, announced that sustainability will be its new investment standard: In particular, given the potential for better outcomes from sustainable investing, BlackRock is incorporating sustainability into risk management, portfolio construction, new product formulation, and engagement with companies. In addition, BlackRock is increasingly working to include certain ESG key performance indicators in its engagement policies, providing clarity on its expectations for companies.

It is worth noting here that risk assessment is one of the main reasons that push companies to undertake ESG analysis. But it is not the only reason. ESG analysis proves to be further useful in seizing, in advance of the wider investor audience, investment opportunities in those companies committed to improving their E, S or G profiles.

Before clarifying the main scientific studies that have debated the role of ESG in the dynamics of a company's performance, it is worth dwelling on one aspect: the integration of ESG criteria is now central also in the evaluation of business risk.

ESG performance provides valuable information on a company's riskiness, complementary to traditional financial indicators. A number of studies show that companies with poor ESG performance have higher risk profiles on various fronts, from credit to operational and reputational risk. In the credit sector, for example, it has been documented that companies with low ESG ratings tend to have a higher Probability of Default (PD) and obtain worse credit ratings from agencies. This is since poor ESG policies increase the probability of adverse events—legal disputes, environmental sanctions, strikes, consumer boycotts, governance scandals—which can result in financial losses and difficulties in repaying debts. Banks and creditors incorporate these risks into rates: empirical evidence shows that companies that are less virtuous on the ESG front pay higher spreads on loans (all other things being equal), reflecting a higher perceived credit risk. A recent study on the Asian market confirms that low ESG scores lead to significant increases in the cost of debt in the following year, thus signalling higher credit risk (Rong & Kim, 2024): "*Extensive media coverage can mitigate (but not eliminate) this effect, acting as an external monitoring mechanism*". Conversely, companies with strong ESG practices are considered more reliable partners: they have better credit ratings (less likely to default) and obtain financing on more favourable terms. Some research indicates that, among the various pillars, the S factor (practices towards employees, communities, customers) can have the most marked positive impact on creditworthiness, as it reduces the risk of conflicts with stakeholders and reputational damage, while very expensive E investments sometimes show mixed effects in the short term. In any case, the direction is clear: high ESG = lower business risk over medium-long term horizons.

Even on the equity and operational risk side, ESG plays an increasingly recognised role. Companies with high ESG standards tend to experience lower volatility in earnings and stock prices, thanks to better management of material risks (e.g.

adaptation to regulatory changes, prevention of environmental accidents or ethical scandals). International evidence shows that portfolios composed of stocks with high ESG ratings experience lower drawdowns during phases of market stress. For example, during the 2008–2009 financial crisis, US companies with a strong ESG orientation outperformed others, better preserving their value (Gianfrate et al., 2024).

Similarly, in the initial impact of the Covid-19 pandemic, studies focused on the American market report that ESG stocks performed better than the market (Lins et al., 2017; Albuquerque et al., 2020). This resilience is often attributed to the social capital built by sustainable companies: relationships of trust with employees, suppliers and customers that translate into cohesion and support in difficult times. In situations of sudden exogenous shocks, the ability to protect workers' well-being and adapt processes—aspects measured by ESG scores—becomes particularly crucial to avoid disruptions and restart quickly. However, it should be noted that resilience is not uniform in every context: a global study of 63 countries found that during the pandemic crisis, companies with high ESG ratings outperformed in North America (equities, bonds and CDS), while in other regions this effect was less pronounced or null (Gianfrate et al., 2024). This suggests that the protective effect of ESG practices may depend on local institutional and market factors (e.g. less national attention to sustainability, less effective welfare systems). In markets such as North America, with fewer public protections, the attribution of value to sustainable companies in times of crisis has been more evident, making these stocks true rainy day assets for investors. In other words, sustainable companies have shown greater financial resilience where sustainability was not yet the widespread norm, while in contexts already strongly oriented towards ESG, the relative advantage in crisis has been diluted.

In conclusion, ESG has become a key factor for a more comprehensive assessment of corporate risk. Integrating ESG criteria into the risk analysis process allows us to capture aspects that are otherwise overlooked but potentially disruptive for business continuity (such as climate risks, social tensions, governance deficiencies). It is no coincidence that financial supervisory authorities are pushing for the integration of ESG factors into credit and management risk models: sustainability is increasingly regarded as the new frontier of creditworthiness analysis and an element to be considered to ensure lasting competitive advantages. A company with high ESG quality is now perceived by investors as less risky and more resilient, a condition that can positively impact its market value. Therefore, proactive management of ESG risks is not only an ethical imperative but also a fundamental pillar for financial stability and long-term value creation.

Academic literature on ESG issues appears to agree with what has been argued so far. Having arisen even before the outbreak of the Covid-19 crisis, the ESG phenomenon has been the subject of multiple scientific studies whose convergent results allow us to state the following:

- *The application of ESG criteria has a positive impact on investment returns in individual stocks.* The main empirical evidence shows that the stock performance of the most virtuous companies in terms of environmental, social and governance

is better than that shown by less virtuous competitors (Eccles et al., 2014; Waddock & Graves, 1997; Dimson et al., 2015; Auer & Schuhmacher, 2016; Khan et al., 2016; Chelawat & Trivedi, 2016).
- *The application of ESG criteria has benefits on the risk profiling and performance of a portfolio*. Empirical data shows that integrating ESG criteria into the investment process and investing in companies with high ESG scores can contribute to better performance (Derwall et al., 2005; Borgers et al., 2015; Tamimi & Sebastianelli, 2017; van Duuren et al., 2016).
- *ESG integration mitigates risk*. Studies show that ESG analysis generates similar returns to non-ESG investments, but at the same time mitigates their risk (Eccles et al., 2014; Morgan Stanley, 2015; Verheyden et al., 2016).

Leaving behind a complicated, not to say tragic, 2022, a future marked by sustainability is taking shape. A future that will see the impact of climate change rise to the main area of risk E, together with the Covid-19 phenomenon, whose social implications will significantly and lastingly impact the area of risk S.

Furthermore, the impact of the controversial Brexit, as well as the uncertainties about the possibility of keeping Covid-19 at bay, will see companies further committed to incorporating the G criteria into their supply chain management.

ESG advocates highlight how the disclosure of ESG information is a critical factor and a determining factor in a company's ability to create value.

The reasons are to be found in the effects arising from ESG reporting (Eccles & Saltzman, 2011; Eccles et al., 2015):

- greater transparency;
- improving internal and external decision-making processes;
- safeguarding and strengthening financial stability.

Several studies, mostly recent, show the significant role of sustainability reports in promoting transparency, as well as the evident implications on corporate finance (Jensen & Berg, 2012; Adams, 2017). Steyn's study (2014) shows the positive impact of sustainability reporting on business and on the company's ability to improve its financial performance. Furthermore, numerous scientific studies indicate that the disclosure of ESG data leads the company to reputational benefits, as well as significant competitive advantages (Gardberg & Fombrun, 2006; Lee Brown et al., 2009; Simnett et al., 2009).

The incessant process of internationalisation and globalisation of markets, the widespread international economic integration and the dimensional growth of companies have pushed many managers and corporate stakeholders not to underestimate and to consider critical and relevant the disclosure of information on corporate governance (Singh & Gaur, 2009). The disclosure of corporate governance data proves to be an effective response to the prevailing institutional environment and influences the decision-making process of stakeholders (Singh et al., 2017).

In the literature, there is some empirical research that highlights how companies that adapt their decision-making processes in relation to the evolution of the external environment are those best able to mitigate the erosive effects on corporate profitability (Seth & Mahenthiran, 2022).

On the other hand, other authors, in contrast with those who affirm the existence of a positive relationship between ESG and corporate financial performance, deny the existence of a positive relationship, as, to date, it is not supported by solid empirical evidence (Larcker et al., 2022).

Chapter 3
ESG Metrics, Indices and Financial Instruments

3.1 ESG Data Providers and the Construction of Sustainability Indicators

ESG indicators are disseminated and shared by sound professional practices, by some data providers (ESG rating providers) and suppliers of specific reports for the measurement and evaluation of the ESG performance of companies allocated to their portfolios. Among the most important, there are:

- Bloomberg ESG Data Service,[1]
- Corporate Knights Global 100,[2]
- Dow Jones Sustainability Index (DJSI),[3]
- ISS—Institutional Shareholder Services,[4]

[1] Bloomberg ESG provides ESG data on over 15,000 global companies, covering approximately 94% of the world's market capitalisation, with verified indicators and scores based exclusively on disclosed data, accessible via Bloomberg Terminal and data licenses. The data is collected from official sources such as financial statements, sustainability reports, proxy statements and questionnaire responses, and is subjected to rigorous quality controls to ensure its reliability. Over 900 ESG indicators can be accessed through the platform, which include aspects such as greenhouse gas emissions (Scope 1, 2 and 3), natural resource management, board diversity, labour policies and much more.

[2] The Corporate Knights Global 100 is an annual ranking of the 100 most sustainable companies in the world, based on 25 ESG indicators and published by the Canadian company Corporate Knights.

[3] The DJSI is a global index that selects the 10% of the most sustainable companies by sector, based on the annual S&P Global Corporate Sustainability Assessment (CSA).

[4] ISS is a leading global provider of corporate governance and responsible investment solutions, providing ESG data, voting advice and analytics to institutional investors and corporates.

A. Quintiliani, *ESG and Value Creation from a Corporate Finance Perspective*, Sustainable Finance, https://doi.org/10.1007/978-3-032-03147-1_3

- MSCI ESG Research,[5]
- RepRisk,[6]
- Sustainalytics Company ESG Reports,[7]
- Thomson Reuters ESG Research Data,[8]
- S&P Global ESG Scores,[9]
- Morningstar Sustainalytics,[10]
- LSEG ESG Score,[11]
- Inrate AG,[12]
- Arabesque,[13]
- Truvalue Labs.[14]

---

[5] MSCI ESG Research provides ESG assessments and analyses on thousands of companies and financial instruments, helping institutional investors integrate sustainability risks and opportunities into their decision-making processes.

[6] RepRisk, based in Zurich, monitors more than 200,000 companies globally, both public and private, assessing their exposure to ESG and corporate conduct risks daily through the RepRisk Index (RRI), which measures reputational risk related to ESG factors. RepRisk methodically and daily assesses risks, allegations and criticisms on issues such as environmental degradation, human rights abuse, child labour, forced labour, fraud and corruption that can impact an organisation's reputation, financial viability or lead to critical compliance and regulatory issues.

[7] Sustainalytics provides ESG Risk Ratings for over 16,000 global companies, assessing their exposure to environmental, social and governance risks and their ability to manage those risks, with a final score that reflects the residual unmanaged risk and potential impact on financial performance.

[8] Thomson Reuters ESG Research Data provides ESG data and scores on over 7000 global companies, based on over 400 indicators and 70 analytical metrics, with an increasing focus on material risks, corporate governance and regulatory integration into decision-making processes.

[9] The S&P Global ESG Score measures a company's performance and management of material ESG risks and opportunities, based on corporate disclosures, stakeholder and media analysis, analytical models and the Corporate Sustainability Assessment (CSA), which includes 62 sector-specific questionnaires.

[10] Morningstar, through Sustainalytics, provides ESG ratings on more than 16,000 companies, integrating quantitative and qualitative data to measure exposure to and management of ESG risks, and develops sustainability indices such as the Morningstar Global Sustainability Index and the Low Carbon Transition Leaders Index.

[11] The LSEG ESG Score, formerly known as the Refinitiv ESG Score, is a rating system developed by Refinitiv (now part of the London Stock Exchange Group) and used by institutional investors to integrate ESG criteria into financial analysis and investment decisions.

[12] Inrate AG, a rating agency based in Zurich, provides ESG Impact Ratings for over 10,000 public companies, based on a proprietary model that assesses the real impact of corporate activities, independently of disclosures alone, with an approach oriented towards dual materiality and the prevention of greenwashing.

[13] Arabesque provides ESG data through the ESG Book platform and the S-Ray tool, which evaluates over 8000 global companies by combining more than 250 ESG indicators with news analysis from over 30,000 sources, offering scores on material performance (ESG Score), compliance with the principles of the Global Compact (GC Score) and climate impact (Temperature Score).

[14] Truvalue Labs is a company that specialises in the analysis of ESG factors through artificial intelligence technologies and alternative data. It uses NLP and machine learning algorithms to monitor in real time unconventional public sources, such as news, social media and NGO reports, to evalu-

Table 3.1 Key ESG issues

Environmental (E)	Addressing externalities arising from the extraction/exploitation of natural resources Responsibility: Energy use Carbon emissions Water management Raw material procurement Impacts on biodiversity Toxic releases Waste and recycling End-of-life products Circular economy
Social (S)	Employee rights—employee safety Employee engagement Diversity, equity, inclusion (DEI) Institutional relations Industrial relations Community relations Consumer health and privacy Access and convenience of products/services Human rights Corporate philanthropy Anti-competitive practices
Governance (G)	Independence of the board of directors Board of directors' responsibility (shareholder vs. stakeholder) Board of directors diversity Executive compensation systems—pay for performance Structure of the internal council committees Corporate code of conduct: Ethics Transparency Risk management Compliance

Source: Personal processing

The criteria used to evaluate the assumption of environmental (E), social (S) and governance (G) responsibility by companies are divided into three macro categories. Table 3.1 summarises the main key issues related to ESG factors.

In the context of the growing importance attributed to ESG criteria in contemporary corporate finance, the schematisation of the main areas of corporate responsibility in the three ESG pillars—as reported in Table 3.1—represents a useful

ate the actual behaviour of companies. Its approach focuses on the dynamic materiality of ESG themes, anticipating risks and opportunities often undetected by traditional rating systems. The company has extended its methodology to other intangible assets, such as reputation, ethics and innovation. It works with global financial institutions to support more informed and data-driven investment decisions. Its analyses allow a continuous assessment of non-financial performance, useful for sustainable investment strategies. The system developed is intended as a complement, and sometimes alternative, to traditional ESG scores.

practical framework to guide the evaluation of the degree of sustainability and value alignment of corporate strategies. This articulation, although descriptive, takes on a profoundly systemic value if read in terms of materiality and impact on the processes of economic-financial value creation in the medium-long term, as is consistently argued in this volume.

The environmental pillar (E), in its thematic articulation, reflects the main negative externalities associated with the production activities of companies, not only in energy-intensive sectors, but in all those contexts in which the life cycle of the product, the use of resources and emissions directly or indirectly affect natural capital. In this sense, corporate responsibility is expressed on multiple levels: from the management of energy and carbon emissions, which are central parameters in the context of the climate transition, to the management of water, waste, materials and recycling processes, which fall within the principles of the circular economy. The emphasis on responsibility related to the end of product life and biodiversity also reflects a growing systemic awareness of the link between production, consumption and environmental degradation. In this sense, the orientation of the company towards regenerative or circular production models represents not only an ethical and regulatory imperative but also a strategic opportunity for competitive differentiation.

In the social pillar (S), the multiplicity of the dimensions considered highlights the evolution of corporate responsibility from a traditionally internal logic—centred on workers' rights and safety—towards a broader relational paradigm, which includes the active involvement of employees, the promotion of diversity and inclusion, union relations and the relationship with local communities. To these are added emerging responsibilities, such as the protection of consumer health and privacy and the accessibility of products and services, which emphasise the dimension of equity in economic relationships. In this context, companies are called upon to account not only for their impact on human rights along the value chain but also for the positive contribution made to social cohesion and local development. The shift from a defensive vision of social responsibility to a proactive and generative vision is the distinctive feature of sustainability in its mature and integrated meaning.

With regard to the governance pillar (G), the dimensions listed in Table 3.1 focus on the quality of decision-making processes, the transparency of strategic choices and the coherence between the corporate governance structure and the long-term objectives of the company. The independence and diversity of the board of directors, the accountability of directors in protecting stakeholders' interests, managerial remuneration criteria aligned with sustainable performance and the presence of specialised internal board committees are all elements that, in the most recent academic literature, are interpreted as fundamental drivers of credibility and accountability. At the same time, the reference to corporate codes of conduct that promote ethics, risk management, compliance and organisational transparency underlines the value dimension of governance, as a space for formalising principles and internalising behavioural norms.

The interaction between these three areas—environmental, social and governance—is not purely additive, but synergistic. The effectiveness of ESG strategies lies in their ability to generate coherence and integration between environmental objectives, social relations and institutional architecture. In the context of corporate finance oriented towards sustainability, as discussed in this book, the joint reading of these three areas allows for a more precise assessment of the company's ability to operate as a responsible, resilient agent and generator of shared value. Therefore, these are not isolated obligations or mere compliance checklists, but rather interconnected dimensions that inform the entire corporate system, from the definition of the strategy to the measurement of results. In this sense, the table constitutes a useful operational reference for mapping the areas of impact and for building indicators that are not only representative but also materially relevant.

In particular, Table 3.1 shows the broad spectrum of ESG issues. A wide range of criteria that sees the main data providers diverge on the key elements to pay attention to when evaluating ESG performance. This divergence is exacerbated by the sectors in which companies operate and/or their geographical area of operation. For example, the release of toxic substances appears relevant in a company in the steel sector, but is irrelevant when analysing the ESG performance of banks or consulting firms. To further illustrate, employee diversity, equity and inclusion (DEI) appear to be critical values to pay attention to in those companies that operate in multi-ethnic countries, such as the USA, while they appear less relevant in those countries, such as Japan, which are characterised by low ethnic fractionation.

Likewise, over time, we have seen a significant growth in equity and/or bond indices dedicated to ESG issues. These indices, often used as benchmarks or for investment strategies, first emerged in 1990 with the launch of the MSCI KLD 400 Social Index, and over time, in less than three decades, we have seen a clear proliferation of indices (NORDSIP, 2019): "*Over 1000 ESG indices in 2019*".

Below, the most popular stock and bond indices are listed in chronological order of launch (from 1990 to 2021):

- MSCI KLD 400 Social Index—since 2010 "FTSE KLD 400 Social Index" (1990),[15]
- Dow Jones Sustainability Index (1999),[16]
- FTSE4Good Index Series (2001),[17]

---

[15] The MSCI KLD 400 Social Index was launched in 1990 and is designed to help socially conscious investors consider social and environmental factors in their investment decisions. It was founded by Amy Domini of KLD.

[16] The Dow Jones Sustainability Index (DJSI) evaluates and compensates the best-performing companies based on economic, environmental and social criteria. Launched in 1999, it reviews a basket of 2500 companies in different economic sectors, from which it selects the most satisfactory in terms of sustainability. It operates through a strategic partnership between S&P Dow Jones Indices and RobecoSAM, an international investment firm focused on financial sustainability.

[17] The FTSE4Good Index Series is a series of ethical investment stock market indices launched in 2001 by the FTSE Group that track the performance of companies that demonstrate strong environmental, social and governance practices.

- MSCI USA IMI ESG Leaders Index (2001),[18]
- WilderHill Clean Energy Index (2004),[19]
- Barclays MSCI ESG Fixed Income Index Series (2013),[20]
- FTSE ESG Index Series (2016),[21]
- MSCI ESG Index Series (2016),[22]
- Solactive ISS ESG Index Series (2016),[23]
- S&P ESG Index Series (2016),[24]

[18]The MSCI USA IMI ESG Leaders Index is a capitalisation-weighted index that provides exposure to companies with high environmental, social and governance (ESG) performance relative to their sector peers. The MSCI USA IMI ESG Leaders Index is composed of large-, mid and small-capitalisation companies from the U.S. market. The index is designed for investors seeking a broad and diversified sustainability benchmark with a relatively low tracking error relative to the underlying equity market. The index is part of the MSCI ESG Leaders Index series. Constituent selection is based on data from MSCI ESG Research.

[19]The stocks and sectors in the index are based primarily on their importance to clean energy, technological influence and relevance to pollution prevention.

[20]The Barclays MSCI ESG Fixed Income Index, launched in 2013, is a fixed income index series developed by Barclays and MSCI that systematically integrates ESG criteria across three categories: Socially Responsible, Sustainability and ESG Weighted, providing benchmarks for institutional investors with ESG strategies.

[21]The FTSE ESG Index Series, launched in 2016, includes eligible securities from the FTSE Global Equity, Russell 1000® and FTSE All-Share® indices, applying FTSE Russell's proprietary ESG scores to increase exposure to companies with strong ESG practices and reduce exposure to those with weak performance, while maintaining a neutral sector weighting to replicate the risk/return characteristics of the benchmark index.

[22]The MSCI ESG Index Series, launched in 2016, integrates MSCI's proprietary ESG scores into global equity benchmarks, selecting and re-weighting stocks based on their management of material ESG risks and opportunities, with the aim of improving the sustainability profile while maintaining consistency with the financial characteristics of the benchmark index.

[23]The Solactive ISS ESG Index Series comprises global equity indices constructed in collaboration with ISS ESG, selecting stocks compliant with international standards on human rights, environment and governance. The indices apply thematic exclusions, controversy filters and a neutral sector weighting, maintaining risk/return characteristics similar to the reference benchmarks.

[24]The S&P ESG Index is a family of market indices designed to provide investors with exposure to stocks with strong environmental, social and governance credentials, while maintaining a structure similar to that of traditional indices. It is not a valuation tool, but rather a financial benchmark that reflects the performance of companies selected based on ESG criteria. The series includes sustainable versions of well-known indices such as the S&P 500, the S&P Europe 350, the S&P Global 1200 and the S&P/Topix 150, thus covering regional and global markets. The selection and weighting of companies within these indices is based on ESG scores provided by RobecoSAM, through a methodology called Smart ESG. This methodology focuses on the financial materiality of ESG factors, i.e., it only assesses those that have a material impact on the economic performance of companies, avoiding biases related to company size or industry sector. In this way, the index maintains a balanced and representative composition of the market, but with a higher sustainability profile. The main objective of these indices is to allow investors to integrate ESG criteria into their portfolios without straying too far from the performance of traditional indices. They are designed for the smart beta market, i.e. investment strategies that combine the simplicity of passive management with the effectiveness of selective criteria. In short, the S&P ESG Index Series represents an

- CAC SBT 1.5 (2023),[25]
- S&P 500 ESG Index (2019).[26]

3.2 ESG Indices and their Implications for Capital Markets

In the current landscape of sustainable finance, ESG indices have taken on an increasingly important role, not only as tools for measuring the non-financial performance of companies, but also as operational devices capable of directly influencing the allocation behaviour of investors, the financing costs of companies and access to capital markets. In a context marked by the progressive internalisation of ESG criteria by economic and institutional actors, the analysis of ESG indices offers a privileged lens for understanding how sustainability is now a structural dimension of financial performance and no longer an accessory or reputational element.

Unlike simple ESG ratings, built for evaluation and prospective purposes by specialised agencies, ESG indices configure synthetic benchmarks that collect baskets of securities selected based on specific sustainability criteria. These indices can assume different configurations depending on the construction methodology, the information sources used, the industrial sector and the geographical area of reference, but they all have a common characteristic: they operate a selection that is based not only on financial variables, but on qualitative and quantitative variables attributable to environmental sustainability, social responsibility and the solidity of corporate governance.

The first internationally recognised ESG index was introduced in 1990 and was called the MSCI KLD 400 Social Index. This benchmark, originally designed for the US market, selected companies with good ethical and social profiles, while excluding companies involved in activities considered controversial, such as tobacco, weapons or nuclear energy. Its evolution in the following years, culminating in its transformation into the FTSE KLD 400 Social Index, paved the way for a proliferation of ESG indices, each with its methodological peculiarities and application purposes.

As previously mentioned, the number of ESG indices has increased exponentially over the last 20 years. According to estimates by Nordic Sustainable Investments (NORDSIP, 2019), in 2019 alone, there were more than a thousand active ESG indices globally. This multiplication reflects a dual dynamic: on the one

evolution of market indices, adapted to the needs of a financial world that is increasingly attentive to sustainability.

[25] This index aims to group 40 stocks whose emissions reduction targets are recognised as being in line with the Paris Agreement, to pursue efforts to limit the temperature increase to 1.5 °C.

[26] The S&P500 ESG Index targets 75% of the market capitalisation of the traditional S&P 500. The index provides diversification and has a profile that is closely aligned with that of the US large-cap market. Companies included in the index are selected based on the S&P DJI ESG scoring methodology provided by Sam.

hand, the growing demand from institutional investors for financial instruments consistent with responsible investment strategies; on the other, the attempt by the main data providers and benchmark suppliers to monitor a rapidly expanding and high-margin market.

Functionally, ESG indices play multiple roles. First, they act as screening and selection tools for investors who wish to build portfolios that comply with certain sustainability standards. Second, they serve as benchmarks for measuring the relative performance of ESG-oriented funds and investment strategies. Third, they indirectly put pressure on listed companies to improve their environmental, social and governance practices to obtain or maintain index inclusion.

The presence or absence of a company in a relevant ESG index can have immediate implications in terms of reputation, access to capital and pricing of securities. Numerous empirical studies, including those of Nofsinger and Varma (2014), Friede et al. (2015) and more recently Avramov et al. (2022), highlight how inclusion in internationally relevant ESG indices is associated with a reduction in the volatility of the stock, an improvement in the risk-adjusted return and, in some cases, a reduction in the cost of capital, whether debt or equity.

The most recent literature highlights how the construction and use of ESG indices are not neutral concerning corporate financial policy choices. Companies that aspire to be included in the main ESG benchmarks are encouraged to adopt more transparent policies on environmental and social matters, to strengthen internal governance and to improve communication with stakeholders. These compliance processes, if on the one hand involve organisational and management costs, on the other hand generate positive externalities in terms of improving the company's risk profile, increasing investor confidence and expanding the pool of potential financiers.

ESG equity indices, such as the Dow Jones Sustainability Index, the FTSE4Good Index, the MSCI USA ESG Leaders Index or the S&P 500 ESG Index, are generally based on best-in-class methodologies that select companies with the best ESG scores within each industry sector. This approach helps avoid sector bias and maintain portfolio diversification while reducing exposure to potentially environmentally or socially harmful corporate practices.

On the bond market side, interest in ESG indices has grown significantly in recent years, in parallel with the development of green bonds, social bonds and sustainability-linked bonds (Cioli et al., 2021). Indices such as the Bloomberg MSCI ESG Fixed Income Index Series or the iBoxx MSCI ESG Screened Indices provide useful tools for building bond portfolios consistent with ESG strategies. Empirical results, including those of Tang and Zhang (2020), suggest that inclusion in such indices can contribute to a reduction in credit spreads, improving the conditions of access to debt by issuing companies.

The regulatory and normative implications of the proliferation of ESG indices are not negligible. The European legislator has recognised the importance of indices as tools for the implementation of sustainable finance policies, introducing disclosure obligations and minimum requirements for climate benchmarks, as provided for by Regulation (EU) 2019/2088 and the subsequent Regulation (EU) 2020/1818. These tools require benchmark providers to justify methodological choices and ensure consistency with the European Union's climate objectives. In this context, ESG indices cease to be simple technical tools and take on a political-institutional value in the transition to a low-carbon economy.

However, there are critical issues. One of the main problems concerns the strong heterogeneity between the methodologies used by the different ESG benchmark providers, which can generate inconsistencies in the classification of securities and in the comparison between indices. The work of Berg, Kölbel and Rigobon (Berg et al., 2022) highlights how the correlation between ESG rating and inclusion in indices is often weak, a sign of a plurality of approaches that are not always transparent. This inhomogeneity can weaken the informative function of ESG indices and make it more complex for investors to orient themselves in the market.

Another problem is the risk of greenwashing, i.e. the possibility that financial products labelled as ESG do not comply with substantial sustainability criteria. In some cases, the simple exclusion of a few companies from an index is presented as ESG transformation, while in substance, the allocation dynamics remain unchanged. The response to this problem has been the development of more rigorous regulatory standards and the adoption, by multilateral bodies such as the International Organization of Securities Commissions (IOSCO), of guidelines aimed at ensuring the transparency and consistency of ESG benchmarks.

Despite these critical issues, ESG indices represent one of the most dynamic and influential tools in the transformation of the global economy towards more sustainable models. Their impact is not limited to financial performance or the reputation of companies, but affects long-term strategic choices, the quality of relationships with stakeholders and the ability of companies to address systemic risks, such as climate change, social inequalities and governance crises.

A first aspect that has emerged strongly from the most recent studies concerns the persistence and comparative performance of ESG indices compared to traditional market indices. A significant contribution in this sense is that of Caporale et al. (2022), who highlight that in emerging markets, ESG indices tend to have a greater persistence in returns compared to conventional benchmarks, suggesting lower information efficiency and therefore greater arbitrage opportunities for informed investors. Differently, in developed markets, this difference tends to be less marked, indicating a better information integration of ESG variables in price dynamics.

The resilience of ESG investments during crisis periods is another crucial topic explored by recent research. An interesting work is that of Broadstock et al. (2021). This study investigates the role of ESG performance during a generalised financial crisis, triggered by the global Covid-19 pandemic. This unique situation provides an opportunity to assess whether investors view ESG performance as a proxy for future stock performance or as a risk mitigation tool. Using an original dataset covering the constituent companies of the Chinese CSI300 index, the results show that: (i) portfolios with high ESG scores generally outperform those with low scores; (ii) ESG performance helps mitigate financial risk during crisis periods; (iii) the role of ESG performance is less relevant in normal market periods, thus confirming its additional importance in crisis contexts. These results are interpreted in the context of ESG investment practices. In parallel, systematic literature offers significant contributions to the debate, as highlighted by Nian and Said (2024). In a systematic review of over 150 studies, the authors demonstrate how ESG performance is generally associated with a reduction in corporate risk and, frequently, with an improvement in financial performance. However, the authors warn against hasty conclusions, given the great methodological and contextual variety present in the different studies analysed.

In line with these observations, the study by Assael et al. (2023) uses advanced machine learning techniques to delve deeper into the relationship between ESG scores and stock returns, finding that the impact of ESG indices on returns varies significantly depending on the sector, company size, and the period analysed. This result highlights the importance of adopting a granular and specific approach when interpreting and using these financial tools.

Complexity and methodological divergences represent further critical issues widely discussed in recent literature. Cesarone et al. (2024), for example, address the problem of discordance between ESG ratings provided by different agencies, proposing a portfolio selection model capable of managing such differences. Their contribution highlights the possibility of integrating ESG ratings strategically, despite their intrinsic variability, to optimise financial and sustainable results.

The empirical analysis conducted by Avramov et al. (2022), which focused on the impact of ESG scores on the performance of stocks included in the S&P 500, adds another layer of complexity to the debate. Contrary to other evidence, this study finds that, at least in the short term, high ESG scores can be associated with slightly lower financial performance compared to traditional benchmarks. This apparent paradox may reflect the initial costs and investments needed to implement robust ESG strategies, the benefits of which typically manifest themselves in the long term.

Another aspect of interest concerns the predictive role of ESG indices, particularly relevant in sectors such as clean energy. A work by Long et al. (2024) shows that ESG stock markets provide useful forecasting signals for the renewable energy sector, highlighting a direct link between the sustainability of financial investments and the dynamics of the energy transition.

Finally, it is relevant to mention the methodological and empirical analysis offered by Berg et al. (2022), who highlight the methodological heterogeneity between ESG rating providers and indices, a problem known as aggregate confusion. This situation generates significant difficulties for investors who wish to use ESG indices to guide their decisions. The authors suggest greater methodological standardisation and more incisive regulation, issues that are also priorities in recent European regulations, such as the SFDR Regulation and the Climate Transition Benchmark.

In conclusion, the review of recent literature on ESG equity indices clearly shows how these instruments have become central to investors and companies, profoundly influencing the financial and operational management of companies. Despite the presence of methodological criticalities and the complexity of the financial effects detected, their ability to orient the market towards greater sustainability emerges strongly, improving both corporate resilience and the overall performance of financial portfolios in the long term. The most recent literature, therefore, invites us to adopt a critical and informed perspective, which takes into account methodological, sectoral and temporal specificities, to best exploit the potential offered by ESG indices.

In the coming years, the evolution of ESG indices will likely be driven by three main factors. First, there will be a push towards methodological standardisation and regulatory harmonisation, with regulators strengthening transparency requirements. Second, there will be an integration of double materiality into selection criteria to capture the financial impacts of ESG aspects, as well as the impacts of companies on the environment and society. Third, there will be a greater focus on sectoral and geographical granularity to avoid overly generic approaches that do not take into account the specifics of the reference contexts.

In conclusion, ESG indices today represent a strategic node for the functioning of financial markets and the governance of the sustainable transition. Their analysis, which combines quantitative tools with qualitative assessments, is becoming a central field of study in contemporary corporate finance. Understanding the logic underlying the construction, selection and use of ESG indices ultimately means understanding the new trajectories of economic value and financial risk in the age of sustainability.

3.3 Sustainable Finance Instruments: From Green Bonds to ESG-Linked Loans and Funds

In addition to constituting fundamental dimensions for the evaluation of non-financial performance, ESG indicators today represent the technical and conceptual cornerstone in the structuring of the main sustainable finance instruments.

The growing sophistication of markets has led to a progressive articulation of financial products that incorporate ESG criteria not only as an ethical constraint but as an objective contractual parameter and, in some cases, conditioning the structure of the return.

In particular, green bonds represent a vehicle through which issuers—public or private—raise capital to finance exclusively projects with positive environmental impacts, such as renewable energy production, energy efficiency or urban reforestation. For the credibility of these instruments, the presence of ex ante (to define eligible projects) and ex post (to measure results) ESG indicators is essential, as required by the ICMA Guidelines (2024) and the European taxonomy of sustainable finance.

Alongside these, Sustainability-Linked Bonds (SLBs) offer an even more dynamic mechanism, as they link the coupon rate to the performance of one or more pre-established, measurable and verifiable ESG indicators. In the event of failure to achieve the objectives, the cost of debt may increase according to predefined contractual formulas.

As Flammer (2021) and Liaw (2020) observe, SLBs represent a significant evolution in the link between sustainability and finance, as they transform the ESG pillar from simple disclosure to a performance lever, directly impacting the structure of financial incentives.

Similarly, Sustainability-Linked Loans (SLLs)—increasingly popular among European companies since 2019—feature bank spreads that vary based on selected environmental or social KPIs, such as reductions in greenhouse gas emissions, improvements in workplace safety or increased gender diversity in senior roles.

However, in order for these tools to truly contribute to the sustainable transformation of companies and not remain mere reputational tools, the ESG indicators used must be materially relevant, verifiable and aligned with medium-long-term industrial strategies.

Several empirical studies, such as that of Sneideriene and Legenzova (2025), highlight that the credibility of such instruments strongly depends on transparency in reporting, independence in external verification and robustness of the metrics adopted, conditions necessary to reduce the risk of greenwashing and strengthen the link between declared sustainability and actual impact. It is precisely the absence of such minimum requirements that represents one of the main vulnerabilities of the current market of ESG-linked products, as highlighted by Dorfleitner et al. (2017), who propose greater regulatory harmonisation and systemic integration between sustainable finance, accounting standards and ESG rating systems.

Central banks and financial supervisory authorities are also assuming an increasingly important role in ensuring the soundness of this market, promoting minimum requirements of transparency, auditability and consistency in the choice of ESG indicators. For example, the European Central Bank has explicitly required supervised banks to include ESG assessments based on solid and materially significant data in their credit-granting processes, highlighting that the integration of ESG factors can no longer be optional, but represents a fiduciary duty towards stakeholders (ECB, 2020).

In light of these considerations, it is clear that sustainable finance tools, particularly ESG-linked products, cannot function without a solid, standardised information system that recognises ESG indicators as structural components of financial evaluation, rather than simple context variables. Their quality, reliability and strategic coherence therefore become determining elements for the overall effectiveness of sustainable finance in promoting real change in corporate practices.

It should be noted that the adoption of ESG tools cannot be reduced to a mere technical innovation or compliance reporting, but is configured as a strategic lever that responds to instances of resilience, competitiveness and alignment with sustainable development objectives. The contribution of Hughes et al. (2021) stands out for the systemic approach with which it describes the relationship between sustainable finance, digitalisation and ESG tools, underlining how the nature of the new tools—from ESG analytics platforms to integrated ratings, from Impact Investing to dynamic evaluation metrics—reflects the urgency of overcoming traditional evaluation schemes focused exclusively on risk-return. From this perspective, ESG tools become enabling devices for a new evaluation paradigm, in which environmental, social and governance dimensions assume a centrality comparable to traditional financial indicators. Recent literature confirms this approach. In particular, the authors analyse how technological innovation is profoundly reshaping ESG evaluation practices, leading to the birth of alternative ESG ratings based on unconventional data sources. Mneimneh et al. (2023) propose an integration between ESG tools and green innovations, showing how environmental metrics can improve thanks to the use of predictive models of emissions. Reshetnikova and Sanak-Kosmowska (2023) highlight the virtuous interaction between global financial security and ESG tools in promoting sustainable development capable of integrating economic resilience and social justice. Furthermore, the same authors underline the importance of information automation and digitalisation as a methodological turning point in sustainability governance. Finally, Sahin et al. (2022) highlight the criticality of the missing pillar in ESG ratings and the importance of information completeness.

Following the considerations set out so far, allow me to examine, in greater detail, some financial instruments whose proceeds are intended to finance economically sustainable projects: (i) Green social and sustainable bonds; (ii) Closed-end funds and open-end funds; (iii) ESG ETFs.

Among the most widespread are certainly green, social and sustainable bonds. Green bonds are relatively new financial instruments, but they have experienced an extraordinary growth rate from 2007 to today. In terms of investment structure, green and social bonds are similar to traditional bonds, but in the allocation of proceeds, issuers must comply with guidelines established according to international principles and shared by market operators. The selection of the project to be financed follows a specific procedure in compliance with a predefined list of categories. In addition, the issuer is required to be fully transparent in communicating and reporting to investors regarding the progress of the financed projects. Furthermore, for financial instruments falling into the green and social category, a series of supports are available to better identify and analyse their characteristics in terms of classification according to international principles, allocation of proceeds and updated reporting on the progress of the projects.

At this point, it seems useful to offer some considerations on funds. Funds can contribute to the financing and long-term sustainable growth of some of the strategic asset classes of the economy. Closed-end funds, thanks in particular to the provision of a fixed amount of capital not repayable by the fund, are among the most appropriate structures to pool capital raised from private investors (retail, professional or institutional) and address certain categories of asset classes in accordance with a determined investment strategy. Some of the most frequent strategies are aligned with the principles of sustainable investment, for the objectives they pursue and, more generally, for the ESG due diligence they apply to the underlying assets.

Let's move on to the third point in the list, namely ESG ETFs. Similar to traditional ETFs (Exchange-Traded Funds), ESG ETFs are investment funds listed on the stock exchange that replicate market indices with respect to ESG criteria. These are instruments that, like closed-end funds and other collective investment vehicles, allow you to exclude from investments oil companies or tobacco multinationals, for example, and instead finance green, ethical or socially useful securities and activities, in line with the United Nations sustainable development objectives set for 2030 (Sustainable Development Goals, SDGs).

Regarding the issue of the convenience of green investment, for some years now, the relationship between finance and sustainability has become increasingly close, so much so that experts in the sector have to think and act in a single direction: to assume sustainability as a strategic lever, as a source of long-term value. Investing in a sustainable way aims to create value for the investor and, at the same time, achieve certain objectives that can satisfy the needs of the present generation, without compromising the possibility that future generations can satisfy their own needs by obtaining social and environmental benefits.

It often happens that common thinking converges towards the association that a sustainable investment portfolio that integrates ESG principles, and especially with a solid ethical filter, corresponds to a lower return, due to a reduced initial investment universe. This myth has been debunked through in-depth studies and empirical research. We can cite the research of Hornuf and Yüksel (2024) in which the entire set of fundamental portfolio managers implements some form of screening, especially with respect to size or liquidity constraints, with the aim of narrowing the basket of investable securities. Scholars clarify that this ESG screening process is capable of integrating additional information that can optimise investment decisions. Therefore, having a very large starting universe available, the application of a strategy that integrates sustainability principles would guarantee the availability of a sufficient number of securities to keep the portfolio diversified and still achieve a higher return. To support this thesis, it is sufficient to compare the long-term trend, from September 2007 to March 2022, of the global sustainable stock index compared to the traditional global stock index: the former is 39.15% higher than the latter. As for return expectations, good corporate management of ESG strategies and risks generally translates into an improvement in operational metrics such as share price, ROI or ROA.

Just to give one example, Larry Fink, chairman and CEO of BlackRock, the world's largest investment firm with over $10 trillion in assets under management, said that the company intends to put sustainability at the centre of its investment decisions in the years to come: "*Our belief, as far as investing goes, is that sustainability and climate-integrated portfolios can provide better risk-adjusted returns to investors*", the CEO said. According to a study by Rockefeller Asset Management in collaboration with the Center for Sustainable Business at New York University Stern School of Business, the institute of the New York university center dedicated to sustainable business, there are five elements determined by the relationship between ESG factors and financial results: (i) Improved financial performance, thanks to sustainable activities, becomes more evident over longer time horizons; (ii) ESG integration, generally as an investment strategy, appears to work better than negative screening approaches; (iii) ESG investments appear to provide downside protection, especially during a social or economic crisis; (iv) Sustainability initiatives in companies seem to favour better financial performance thanks to mediating factors such as better risk management and greater innovation; (v) Studies and industry research indicate that low carbon emissions improve financial performance and the disclosure of ESG activities does not directly determine financial performance. Furthermore, regarding the outlook, the trend seems destined to last and consolidate. In fact, until a few years ago investors have always focused on instruments and sectors that they knew well, for example stocks or bonds, or that they considered safe, such as gold or long-term government bonds, while sustainable investment was seen as a fad or, worse, a luxury and a field to look at only sporadically. I believe this thinking belongs to the past and is destined never to return in the future. Companies that create value for their stakeholders, monitoring externalities, are benefiting from their commitment, and funds that invest in companies with strong ESG policies outperform their benchmarks. From a risk management perspective, supporting these companies is, logically, the right decision to make. This is just the beginning: we are witnessing a major shift in corporate behaviour, especially as hundreds of companies have been forced to publicly reassess their relationships with customers, employees, suppliers and the community at large. Looking ahead, companies that truly value stakeholder relationships will be best placed to emerge stronger from this crisis. Another aspect to consider is that the ESG world was driven, especially at the beginning, by the equity component and the environmental aspect: sustainable products were predominantly equity funds that invested in the environmental world, the so-called green funds, those that invest in companies that deal with renewable energy production, energy transition and so on. Today, the spectrum of product offerings has changed radically and social and governance aspects are also starting to gain relevance, a trend that may strengthen in the future.

Finally, from the point of view of companies, attention to the topic of sustainability has grown rapidly in recent years, also from a regulatory perspective, and, in this context, companies are called upon to answer for the impact of their business and to approach consumption and investment choices that are more oriented towards environmental and social responsibility. As far as Italian companies are concerned, attention to sustainability issues is growing significantly, even if there is still a long way to go in terms of communication, in line with internationally recognised standards regarding the results achieved. Having previously clarified the differences between CSR and ESG, it is worth specifying that CSR is now recognised as an objective with possible economic, as well as social, value to be pursued through targeted actions. In the permanent census of the Italian Istat, CSR has been explored by distinguishing the types of intervention most relevant for workplace well-being and equal opportunities (internal CSR) and for the territory (external CSR), and the presence of a system for evaluating the effectiveness of these interventions by companies (on the more general commitment of companies to sustainability actions). In conclusion, there are numerous advantages and some disadvantages of sustainable investments. Sustainability is a path of improvement to consolidate the competitiveness of companies. The advantages deriving from the implementation of an ESG strategy can be summarised in five factors: (i) Growth of the company's brand reputation; (ii) Facilitated access to sustainable finance that allows companies to increase the monetary and financial value of the company; (iii) Centrality of stakeholders and enhancement of the context in which the company is inserted; (iv) Greater involvement of top management in the process of defining and monitoring ESG results; (v) Furthermore, implementing a sustainability strategy consistent with the business model allows to increase the economic value generated thanks to the increase in revenues, to remain in the supply chain of sustainable companies and the possibility of accessing national and international public tenders. Similarly, the main disadvantage of a sustainable investment lies in the difficulty encountered in analytically evaluating the impact generated by sustainability practices.

In the overall picture of general and growing interest and favour for ESG issues described so far, there is no lack, in reality, of discordant positions both on the analysts' and operators' fronts. The most striking of the former is that of Aswath Damodaran, professor of finance at the Stern School of Business of New York University, where he teaches corporate finance and stock valuation, who represents the most critical voice on ESG scoring. Damodaran, an authoritative and respected voice in the academic and financial landscape, with a long series of publications to his credit, is also one of the most ardent critics of the ESG world, which he did not hesitate to define as a scam that enriches consultants, measurement services and fund managers, while doing almost nothing for the companies and investors it claims to help and even less for society. As for the latter, the news that BlackRock, one of the largest investment management companies in the world, has significantly reduced its commitment to environmental and social issues, approving only 7% of the related shareholder proposals in 2023, is sensational. In short, Larry Fink, the CEO of BlackRock, has made strong statements: he has declared that he no longer intends to use the acronym ESG. According to the CEO, these factors have become double-edged swords that politics has used to orient the market. In reality, the group's position appears to go against the general trend: in the US in this first part of 2025, there have already been voted 340 proposals in the ESG direction, a record considering the 300 approved in all of 2022. At the same time, according to a recent study by KPMG (2023), more than half (53%) of investors have cancelled M&A operations due to the results of ESG due diligence. Furthermore, if ESG deficiencies do not succeed in cancelling a deal, they can still depress the sale price. Four in ten US respondents (42%) said that ESG due diligence results led to reductions in purchase prices. Exceptional ESG performance can, on the other hand, increase prices. Over 60% of investors said they were willing to pay a premium for companies with a high degree of ESG maturity and in line with their priorities. Of these, over a third said that the premium could exceed 5%. Therefore, even with authoritative opinions and significant discordant positioning, the market also seems to be increasingly oriented in the direction imposed by policy makers in the ESG area, a trend that However, it will have to find its own correct adjustment over time, also thanks to the changes and additions that operational experience will determine.

Chapter 4
ESG Factors and Firm Financial Performance

4.1 From Ethical Commitment to Financial Impact: An Overview

ESG criteria are considered non-financial performance indicators and are used to identify good corporate practices in terms of ethics, social responsibility and corporate governance. In other words, the ESG theme identifies a parameter for evaluating the performance of companies, together with traditional economic parameters, by the financial world (banks and institutional investors) but also by the so-called supply chain leaders, i.e. companies that place themselves at the head of the supply chain by asking their suppliers to adapt their management models to the principles of sustainable development.

In the field of ESG investments, a key concept emerges, among others, namely ESG integration. In this regard, the UN Principles (Principles for Responsible Investment—PRI) define it as follows: "*The explicit and systematic inclusion of ESG factors in investment analysis and decisions*". In other words, ESG integration recalls the application of ESG factors in traditional financial analysis.

ESG, or ESG factors, are increasingly receiving attention from finance professionals, individual investors, institutional investors and professional fund buyers.

To be recognised and labelled as socially responsible, recent years have seen an ever-increasing number of companies engaging in ESG activities. At the same time, investment initiatives by an increasing number of asset managers, pension funds and institutional investors have been undertaken after assessing the ESG activities of the target companies.

The financial industry and academia reveal investors' now strong awareness of the existence of an intimate link between ESG and corporate finance.

A survey conducted by Accenture in 2021 on a large panel of institutional and non-institutional investors highlights how 72% of respondents incorporate ESG principles into their investment approaches and decision-making processes.

A. Quintiliani, *ESG and Value Creation from a Corporate Finance Perspective*, Sustainable Finance, https://doi.org/10.1007/978-3-032-03147-1_4

Among those already implementing ESG strategies, 46% believe that this analysis is as important to their investment process as traditional fundamental analysis. 90% of respondents believe that ESG-integrated portfolios will outperform the market and non-ESG-integrated portfolios going forward. In addition, some institutional respondents see the potential for enormous challenges that will arise from ESG assessments.

Here, supported by existing literature, we intend to clarify how ESG practices can play a key role in generating value for a company. In other words, the aim is to clarify how ESG performance (ESG scores) can be reflected in the elements that constitute the value of a company: cost of equity, cost of debt, return and risk.

Before clarifying these aspects, it seems appropriate to take a step back with the aim of better clarifying the state of ESG research in corporate finance studies (Table 4.1). In this regard, it seems useful to recall some valuable and interesting evidence that emerged from Gallucci's study (2023) on the occasion of the ADEIMF (Associazione Docenti Economia Intermediari Mercati Finanziari) Winter Conference 2023.

The first investigative perspective (characteristics of the reference markets) has seen multiple authors investigate the impact of ESG performance on stock returns, depending on the countries/markets of origin/operations of listed companies. This is a broad and complex field of investigation, within which some empirical studies are placed that demonstrate how ESG criteria play a critical, strategic and more significant/impactful role in those companies that, unlike others, operate in low-trust countries/markets where poor compliance with ESG principles, poor measures to protect minority shareholders, poor transparency, widespread corruption, poor workplace safety regulations and low standards of sustainability disclosure prevail (Karolyi et al., 2020; Neukirchen et al., 2023).

The second perspective of investigation (corporate governance) highlights how corporate governance can act as a strategic lever to promote ESG practices in corporate finance. Analysing the key variables, we note that attention to the composition and characteristics of corporate leaders can be decisive in orienting the company towards sustainability objectives. Corporate governance variables such as the active presence of members on the board, gender diversity and the characteristics of the CEO are particularly significant. A more present board, for example, is often associated with more rigorous oversight of ESG issues, which strengthens the company's responsibility and transparency. The presence of women on the board also enriches the quality of governance: studies show that boards with a greater female representation are more inclined to promote inclusive and socially responsible initiatives. The personal characteristics of the CEO, such as age and level of self-confidence, can also influence the propensity towards ESG practices, with younger or more confident CEOs tending to be more willing to adopt sustainability policies. CEO compensation also plays a strategic role: linking compensation to ESG performance can incentivise an alignment of interests between leadership and sustainability, favouring the integration of ESG practices in corporate strategic choices. As for future research paths, these expand the scope of analysis and focus on the sociodemographic characteristics of corporate leaders as possible determinants of

Table 4.1 The state of ESG research in corporate finance studies

Investigation perspective	Current research question	Future research paths	References
Characteristics of the reference markets	What are the institutional characteristics of a country and the type of effect generated on the ESG activities of a company?	Examine how key institutional stakeholders can influence a company's sensitivity to sustainability (ESG) issues—customers, workers, politics, and other relevant actors	Dicko et al. (2020) Dai et al. (2021) Nekhili et al. (2021)
Corporate Governance	What are the corporate governance variables and the type of effect they generate on a company's ESG activities?: Presence of directors on boards Presence of women on boards CEO characteristics: age, confidence level CEO compensation	To examine how personal and sociodemographic characteristics of corporate leaders can influence a company's sensitivity towards sustainability (ESG) issues (gender, political orientation, integrity, personality, power) Examine the critical mass of different categories of directors Composition and role of the internal council Sustainability Committees Increase the number of independent variables to analyse	Li et al. (2018) Pham and Tran (2020) Dabbebi et al. (2022)
Ownership structures	Do ownership structures influence a company's ESG activities?	Heterogeneity of family ownership, institutional ownership and state ownership can influence a company's sensitivity to sustainability (ESG) issues Engagement	Rees and Rodionova (2015) Aluchna et al. (2022) Thuy et al. (2024)
Risks and cost of capital	Do ESG activities influence the risks a company faces and, ultimately, its cost of capital?	Vast empirical evidence supporting the reduction of business risk and its cost of capital	Gillan et al. (2021)
Performance and value	Do ESG activities influence a company's performance and value?	Examining how a company's ESG activities influence value (Tobin's Q) Increase the number of dependent variables considered	Fatemi et al. (2018) Yu et al. (2018) Wong et al. (2021)
Banking relationship	Does the bank-company relationship influence a company's ESG policies?	What is the mechanism (direct influence, voice, exit) through which banks influence ESG policies? Do sustainability awareness-raising activities in the areas where companies operate influence ESG policies? Does increased transparency in credit assessment processes influence ESG policies?	Houston and Shan (2022) Degryse et al. (2023)

Source: Gallucci (2023)

sustainability orientation. Analysing factors such as gender, political orientation, integrity, personality and power could reveal significant influences on corporate leaders' sensitivity towards ESG issues. Similarly, deepening the influence of the critical mass of different types of directors would allow us to assess whether a certain number of members with ESG orientations can determine an effective change in corporate policy. The composition and role of internal board Sustainability Committees also deserve attention to understand whether and how these committees can strengthen ESG responsibility. Finally, an expansion of the number of independent variables to be examined could enrich the understanding of ESG determinants. References to the literature (Li et al., 2018; Pham & Tran, 2020; Dabbebi et al., 2022) highlight the growing research attention towards the role of governance variables in promoting ESG activities. Deepening these studies can consolidate the theoretical and methodological bases to understand how corporate governance and the profile of leaders influence sustainability performance. Corporate governance, therefore, appears crucial to support the transition towards more sustainable corporate practices, and a better understanding of the dynamics between leadership and ESG orientation could prove crucial for the future of corporate finance.

The third research perspective (ownership structures) offers relevant and still evolving research perspectives. The current research question focuses on the influence of ownership structures on a firm's ESG initiatives, highlighting a link between ownership structure and strategic decisions in the sustainability field. This approach is significant because it allows for the examination of the role of owners as key agents for the integration of ESG practices into corporate policies. The idea of exploring the heterogeneity of ownership structures opens up interesting research avenues. Family ownership, for example, is often linked to a long-term perspective and a greater sense of responsibility towards stakeholders, which could translate into a greater sensitivity to ESG issues. However, these firms may also adopt a conservative approach and limit ESG investments in contexts of uncertainty. Firms with institutional ownership, on the other hand, may be more inclined to integrate ESG practices to respond to pressure from funds and investors that prioritise compliance with sustainability criteria. Finally, state ownership, often linked to public policy objectives, could promote ESG strategies in line with national or international priorities, although in some cases, an instrumental approach could emerge depending on the prevailing political interests. Engagement represents another interesting perspective, as it analyses how different types of ownership interact with firms to influence their ESG choices. Exploring the mechanisms through which owners exert this influence could help clarify the role of ownership dynamics in sustainability pathways. The studies recalled in Table 4.1 (Rees & Rodionova, 2015; Aluchna et al., 2022; Thuy et al., 2024) provide an important basis to build these lines of research. These studies allow us to understand how the intersection between governance, ownership and sustainability is treated in the academic literature, laying the foundations to delve deeper into the mechanisms linking different types of ownership to ESG activities.

4.2 ESG Performance and Profitability: Evidence and Interpretations

In the literature, there is an intense academic literature on the causal links between ESG performance and firm financial performance. However, the non-uniqueness of the data and the extreme variety of approaches to reporting ESG performance are reflected in the scientific results, which are quite discordant (Li et al., 2018).

Why should ESG practices improve financial performance? In an attempt to answer this question, a widely accepted and widespread theory has emerged that links better financial performance to a reduction in the cost of capital. The supporters of this theory highlight how the higher costs incurred to adapt to ESG criteria are reflected, virtuously and to the same extent, in the reduction of the cost of capital. In this regard, Mackey et al. (2007) clarify that behaviour that can be qualified as socially responsible can be associated with a product sold by companies to investors. But does this product create value for the company? In line with the cost of capital theory, the study conducted by El Ghoul et al. (2011) highlights a negative correlation: the higher the ESG rating, the lower the cost of capital.

Albuquerque et al. (2012) believe that ESG is a product that can be classified as strategic and, therefore, a source of greater profits for the company.

Particularly interesting is the study conducted by Sharfman and Fernando (2008), which demonstrates how the disclosure of extra-accounting information and ESG reports is are valid tool to support the management and control of corporate risks. Therefore, companies with higher ESG ratings are considered less risky.

Over time, numerous theoretical and empirical studies have highlighted the correlation between ESG performance and corporate financial performance.

Friedman, Mackey and Rodgers (2005) find a negative correlation. This scientific evidence is consistent with neoclassical economic theory, which states that companies' ESG conduct causes competitive disadvantages and an unjustifiable increase in operating costs (Aupperle et al., 1985; Mcwilliams & Siegel, 1997; Jensen, 2002).

In line with those who highlight the negative correlation, the statements of Perrini et al. (2011) clarify how ESG issues should be addressed only by non-profit organisations.

In particular, supporters of shareholder theory highlight how ESG practices are irreconcilable with the company's objectives aimed at maximising the wealth of its shareholders (Jensen, 2002).

On the contrary, very quickly we are witnessing a fruitful academic literature that sees the supporters of the stakeholder theory supporting the positive correlation between ESG performance and corporate financial performance (Donaldson & Preston, 1995). On the basis of this theory, companies could improve their profitability by satisfying stakeholders' expectations (Freeman, 2010).

Other studies clarify that financial performance is subordinated to the company's ability to manage relationships with key stakeholders (Waddock & Graves, 1997).

For many, ESG performance is comparable to an intangible that can lead the company to a more efficient and strategic management of available resources.

The disclosure of ESG information prompts companies to pay greater attention to environmental matters, employee well-being, shareholder interests, and broader societal concerns. In other words, companies are obliged to behave responsibly under ESG policies. For example, a company oriented to compliance with laws and ethical standards in waste management is less exposed to the risk of environmental sanctions.

Furthermore, compliance with ESG principles allows the company to improve its market positioning. From this, it builds a stable and sustainable relationship with its stakeholders.

For greater clarity, the measures that ensure a safe and comfortable work environment, serene and favourable to interpersonal relationships, have a positive impact on organisational well-being and offer clear advantages in terms of productivity, turnover and customer satisfaction. Satisfied employees are much more productive than dissatisfied ones. In particular, their satisfaction has a positive impact on customer satisfaction with clear benefits in terms of image return and loyalty, the latter being critical factors that determine the company's ability to successfully acquire new market shares or segments (Verheyden et al., 2016). Similarly, actions useful for ensuring a correct implementation of corporate governance practices prove suitable for protecting the interests of minority shareholders.

In summary, ESG disclosure enables investors to (Murata & Hamori, 2021):

- streamline the securities portfolio,
- mitigate information asymmetries between insiders and outsiders,
- discourage moral hazard,
- mitigate the risk of adverse selection of corporate stakeholders,
- increase enterprise value.

Several authors clarify that companies that are inclined to manage/optimise relationships with their stakeholders, and that achieve the best ESG performance, are characterised by a high reputational profile, with evident benefits in terms of competitive advantages and ability to protect themselves from risks during times of crisis (Barnett & Salomon, 2012; Do, 2022). Therefore, the company manager is encouraged to improve ESG performance, as the latter:

- increase the size of the company's reputational capital;
- consolidate the relationship between the organisation and stakeholders and, lastly;
- generate business value.

In particular, noteworthy is the empirical evidence emerging from the study by Fatemi, Glaum, and Kaiser (Fatemi et al., 2018); scholars state that the disclosure and communication of ESG information positively impact the financial and market value of a company.

In the literature, there are two lines of study that investigate the relationship between ESG and the risk/return ratio.

A first line of research, the richest in terms of scientific contributions, focuses on the study of the relationship between ESG and corporate profitability (Alshehhi et al., 2018; Fatemi et al., 2018; Yoon et al., 2018; Dalal & Thaker, 2019; Bhaskaran et al., 2020; Abdi et al., 2022); differently from the first, the second line of study analyses the relationship between ESG and the cost of capital (cost of equity and cost of debt).

Literature focused only on performance appears misleading, as ESG factors can affect both risks and returns of companies. Some studies show that corporate investments in sustainability can be crucial and strategic in the attempt of corporate governance to mitigate the cost of capital, which includes the cost of debt, to be associated with creditworthiness, and the cost of equity.

Table 4.2 lists the main studies that have analysed the relationship between ESG and income performance; in this, there are: the number of sampled companies and their country of origin; the survey period; the number of observations; the ESG data providers; the regression method; the correlation, positive (+), negative (−) or not significant (ns), with income performance; the pillars.

Without prejudice to the considerations presented so far, it is clear that in the literature, there is a growing consensus on the positive link between ESG practices and corporate financial performance. Numerous recent meta-analyses highlight that most studies find a non-negative (often positive) relationship between high ESG scores and corporate results. For example, an analysis of over 1000 research studies (2015–2020) shows that in 58% of cases, ESG is associated with better operating metrics (ROE, ROA, share price), with only 8% of negative results. In parallel, approximately 59% of investment studies find similar or superior performance for sustainable portfolios compared to traditional ones (Khan et al., 2021). These findings indicate that integrating ESG factors into corporate strategy can translate into greater operational efficiency and more robust stock returns. Indeed, focusing on material ESG aspects—relevant to the sector—tends to be indicative of high-quality management oriented towards long-term value creation. However, there is no lack of heterogeneous results and heated debates. The difficulties in univocally measuring ESG performance (divergent ratings between providers, non-standardised metrics) may explain part of the variability of the results. Some older studies from a shareholder theory perspective supported a negative correlation, arguing that ESG initiatives increase costs and reduce competitiveness (Jensen, 2002). Even today, some authors question the existence of a clear and robust link, believing that the empirical evidence supporting an ESG premium is not yet completely convincing (Larcker et al., 2022). However, the prevailing orientation of the most recent studies supports the stakeholder-oriented view, according to which investing in sustainability can generate net economic benefits for the company in the medium-long term. For example, during the Covid-19 crisis, it emerged that companies with a high focus on human capital and supply chain (social and governance ESG pillars) suffered lower stock market declines: one study found an excess return of +1.4%/+2.7% for companies with a high crisis response score compared to others in the first quarter of 2020 (Cheema-Fox et al., 2020). This suggests that managing ESG factors well can contribute to greater operational and financial resilience, protecting

Table 4.2 Doctrinal guidelines: correlation between ESG scores (or ESG ratings) and income performance

Authors	Sample	Period	Obs.	Data provider	Regression model	Positive relationship (+) with income performance?	Pillar
Velte (2017)	Listed companies (German Prime Standard—DAX30, TecDAX, MDAX)	2010–2014	412	Thomson Reuters	Multivariate	(+) ROA	E S G
Li et al. (2018)	350 listed companies (FTSE UK)	2004–2013	548	Bloomberg Boardex	Multivariate	(+) ROA	E S G
Atan et al. (2018)	54 companies (Malaysia)	2010–2013	162	Bloomberg	Multiple Linear. Least Squares Method—OLS	(+) ROE	E S G
Bellavite Pellegrini et al. (2019)	182 listed companies (oil and gas industry)	2002–2018	3094	Thomson Reuters	Easton—with fixed effects	(−) ROA	E S G
De Lucia et al. (2020)	1038 companies (EU countries)	2018–2019	5000	Thomson Reuters	Machine learning and Logit	(+) ROA (+) ROE	E S G
Al Amosh et al. (2022)	124 companies (Levant countries): Jordan, Palestine, Syria, and Lebanon	2012–2019	883	Bloomberg	Data Panel	(+) ROA, Pillar G (+) ROE, Pillars E, S	E S G

Source: Personal processing

corporate results even in adverse contexts. In short, although there are exceptions, the most recent empirical evidence confirms that ESG criteria, when integrated into strategic decisions, tend to align the interests of the company with those of stakeholders, improving both economic performance and the sustainability of the business in the long term.

Another element of concern is represented by the uncertainty in ESG ratings. According to Avramov et al. (2022), this uncertainty increases the risk premium required by investors, penalising companies that are less transparent and with discordant ratings.

4.3 ESG and Creditworthiness: Rating Dynamics and Risk Metrics

Some studies show that ESG factors have a positive influence on credit rating (Ashbaugh-Skaife et al., 2006; Bouslah et al., 2013). In particular, the S factor shows the most significant positive impact on creditworthiness (Bauer et al., 2009; Jo & Harjoto, 2011), while the E variable surprisingly shows negative effects (Sharfman & Fernando, 2008).

With regard to the bank-firm relationship, several studies highlight a negative correlation between the ESG performance of the potential borrower and credit risk. These studies demonstrate that poor ESG performance can ultimately translate into greater credit risk. The causes are to be found in the increased probability of conflicts, costly and protracted over time, that would arise between companies with lower ESG performance and their stakeholders. Conflict risks, such as negative publicity, consumer boycotts, litigation, which reverberate on the Probability of Default (PD) and which, therefore, banks incorporate into their pricing and financing contracts (Sharfman & Fernando, 2008; Goss & Roberts, 2011; Chava, 2014; Hasan et al., 2017; Hauptmann, 2017).

Further studies find a negative relationship between ESG investments and credit rating. This relationship is justified for quite different reasons:

- first, large investments in ESG may cause agency conflicts between company managers (agents), who could benefit from investment policies, and shareholders (principals), who would be forced to bear the higher costs associated with large investments (Goss & Roberts, 2011);
- secondly, achieving a high level of ESG performance, as it is anchored to the company's ability to maintain stable relationships with its stakeholders, requires the incurring of additional costs, as well as being a source of increased fixed costs (Perez-Batres et al., 2012).

Regardless of what has been stated so far, consistently with what has been argued by the European supervisory authorities EBA and ESMA, which will be explored further below, it is reasonable to think that the integration of ESG criteria in the creditworthiness assessment process is the new frontier of sustainable finance. Therefore, sustainability is increasingly becoming a central element to be integrated into risk analyses and to be considered to create value and competitive advantage.

Given that the level of ESG performance of a company can impact the Probability of Default (Yadav & Asongu, 2025), the exposure of a counterparty or loan to ESG factors can be captured through an ad hoc rating system (sometimes called sustainability rating or ESG rating). The sustainability rating is a synthetic judgment that certifies the solidity of an issuer, a security or a fund from the point of view of environmental, social and governance aspects.

It is worth noting that the need to incorporate ESG factors into creditworthiness analyses has been called for on several occasions by the EBA and ESMA, called upon to provide the European Commission with a technical opinion on sustainability factors and risks.

Since 2019, the EBA (European Banking Authority), together with the ESMA (European Securities and Markets Authority) and the European Commission, have published several regulations on how to integrate ESG into credit ratings.

In particular, after a long period of discussion with the various stakeholders, the EBA published on 29 May 2020 the new Guidelines on Loan Origination and Monitoring (LOM), which aim to improve the practices, governance and processes adopted by financial institutions when granting and monitoring credit.

The Guidelines on Loan Origination and Monitoring (EBA LOM), which also came into force in Italy on 1 July 2021, provide the banking system with a new set of documents and information to request from the world of Small and Medium Enterprises (SMEs), functional to the assessment of creditworthiness also from an environmental and social point of view. In short, the EBA clarifies that all companies, including small and medium-sized ones, in the coming years will have to take into account not only financial aspects but also environmental, social and governance aspects to access new capital.

In particular, the Guidelines provide a method of determining pricing in a new risk-based perspective, proposing a series of minimum elements that banking institutions should consider when granting new financing; among these, the EBA requires that environmental and social aspects are taken into account and how much the company can look to the future in a sustainable perspective.

In terms of governance and credit granting, the EBA takes up the indications received from the European Commission, placing strong emphasis on the need for banking institutions to incorporate an eco-sustainable culture into their internal processes, managing the risks associated with the non-sustainability of the business and the activities of the entrusted customers.

The Guidelines also introduce strong indications so that the banking system integrates the assessment of the creditworthiness of economic entities with ESG components, evaluating their exposure and the related risk. At the same time, the EBA invites the banks themselves to implement credit products that are intended for projects that intend to pursue eco-sustainable investments or for entities that want to improve their virtuous path of eco-sustainability of their business (green finance).

Based on the indications provided by the EBA, banking institutions have been invited, especially in a first phase, to complement the traditional rating of an SME with ESG elements that are attributable to its sector of belonging or the related supply chains, but there are already several banks that are integrating their qualitative analysis modules with information to be asked of SMEs on the subject of eco-sustainability, to define their exposure to ESG factors in an even more specific manner (Nurlanovich, 2025).

For greater clarity regarding the inclusion of ESG issues in the creditworthiness assessment, it is considered necessary to recall the provisions of the new EBA Guidelines on the granting and monitoring of credit:

Section 4.3—Credit Risk Policies and Procedures.

Paragraph 56—Institutions should incorporate ESG factors and the risks associated with them into their credit risk appetite, risk management policies and credit risk policies and procedures, taking a holistic approach.

Paragraph 57—Institutions should take into account the risks associated with ESG factors for the financial condition of borrowers, and in particular the potential impact of environmental factors and climate change, in their credit risk appetite and credit risk policies and procedures. Climate change risks for the financial performance of customers may materialise mainly in the form of physical risks, such as those arising from the tangible effects of climate change, including liability risks for contributing to climate change itself, or transition risks, such as those arising from the transition to a low-carbon and climate-resilient economy. In addition, other risks may arise, such as changes in market and consumer preferences and legal risks, which could affect the performance of the underlying assets.

Section 5.2—Customer Credit Assessment.

Paragraphs 126 and 146—Institutions should assess the client's exposure to ESG factors, in particular environmental factors and the impact on climate change, and the appropriateness of mitigation strategies, as specified by the client. Only for loans to micro and small enterprises is it foreseen that this analysis should be carried out at the client level. However, where appropriate, institutions may also consider carrying out this analysis at the portfolio level.

Paragraphs 127 and 149—In order to identify clients that are exposed, directly or indirectly, to higher risks associated with ESG factors, institutions should consider using heat maps that highlight, for example, the climate and environmental risks of individual economic (sub)sectors in a graph or on a metric. For loans or clients associated with higher ESG risk, a more in-depth analysis of the client's actual business model is required, including a review of current and projected greenhouse gas emissions, the market environment, ESG supervisory requirements for the companies under consideration and the likely impact of ESG regulation on the financial position of the client.

Remembering that the attribution of an ESG rating requires the analysis of Environmental, Social and Governance factors, reading the content of the Guidelines leads to the following considerations:

- The emphasis is placed on the environmental factor, which is the factor for which the most robust correlation analyses with insolvency risk exist.
- The analysis of the environmental factor includes the assessment of the transition risk and the physical risk. At least for SMEs, the physical risk will almost certainly be assessed at the system level, based on information relating to the location of the companies' production facilities and the economic sector to which they belong.
- The methodology for incorporating ESG factors into the credit rating process is not identified. For instance, it remains unclear whether ESG considerations are integrated into pre-existing rating models, whether internal or external ESG

scores are used to adjust model-generated ratings, or whether a distinct approach is adopted altogether.

The last consideration seems to suggest a credit rating system that will see financial intermediaries compete on their ability to incorporate ESG factors into their rating systems.

Looking ahead, Internal Rating Systems (IRSs) will have to incorporate traditional ratings with sustainability ratings. In this perspective, the intangible nature of multiple ESG factors will require banks to build their internal rating systems by drawing on soft information mostly obtained from questionnaires (qualitative and very often from external databases), ESG reports (where present), on-site visits and direct assessments functional to investigate the responsible and sustainable behaviour of counterparties.

Soft information requires an appropriate classification. These can be distinguished into subjective and objective (Quintiliani, 2016).

For clarity, an objective qualitative datum can be identified in the amount of greenhouse gas emissions—GHG GreenHouse Gas, while a subjective qualitative datum can emerge from the statement of GHG emission reduction objectives.

While qualitative data with objective content will be monitored and validated by rating systems in the same way as quantitative variables, qualitative data of a subjective nature will be processed by the bank following a judgmental approach, i.e. treated on the basis of the experiences and knowledge of those who manage the relationship with the customer. It should be noted that internal rating systems must prove suitable for capturing ESG risks, as they impact the known risk parameters—financial risk (EBA, 2021):

- credit (impact on parameters: PD, LGD and EAD),
- market (impact on the value of the financial instruments issued,
- operational (legal and reputational risk),
- liquidity and funding (value of financial assets).

For example, the Probability of Default (PD) may vary as a result of:

- *New regulations in the area of sustainability*, which could reduce the demand for some products and, therefore, cause a drop in turnover.
- *Adverse weather conditions*, such as floods, which could be a source of crisis in particular sectors, such as agriculture.
- *Exposure to companies that violate human rights.*

Exposure at Default (EaD) could decrease if the counterparty is affected by unforeseeable exogenous events, such as environmental disasters.

Loss Given Default (LGD) could be negatively impacted by the loss in value of assets used to secure loans, leading to losses when non-performing exposures are collected.

4.4 ESG and Cost of Capital: Implications for Debt and Equity

Over the last few years, the topic of the role of ESG factors on the processes of creating corporate value has been investigated with the clear intent of understanding the links between ESG variables, ESG performance and aspects strictly connected to the world of corporate finance and capital budgeting, such as the cost of capital and corporate profitability.

International research and analysis have found that improved financial performance in terms of sustainability is linked, for example, to a reduction in the cost of equity for companies and, consequently, in the WACC.

Since WACC is perceived by the investment community as a key indicator in measuring firm value, it is clear that ESG practices must be analysed for their impact on the cost of capital.

For clarity, the cost of capital is one of the most developed areas of corporate finance, particularly following the definitions of Modigliani and Miller, William Sharpe's Capital Asset Pricing Model (CAPM) and the entire financial theory.

In the theories, at least three types of cost of capital are highlighted:

- cost of equity,
- cost of debt;
- weighted average cost of capital (WACC).

The WACC is nothing more than a weighted average of the cost of debt capital and the cost of equity capital, with weighting factors equal to the weights that debt and equity capital assume on the value that emerges from their sum. The formula (Eq. 4.1) greatly simplifies this last concept (Quintiliani, 2008):

$$\text{WACC} = K_e \frac{E}{D+E} + K_D \frac{D}{D+E} \tag{4.1}$$

Where:

K_e = Cost of equity,
K_d = Cost of debt,
D = Debt,
E = Equity.

The cost of debt (K_D) for the company represents the remuneration requested by the external financier as a function of the financial risk perceived by the latter: the greater the risk (for example, the company being financed is highly indebted), the greater the remuneration requested by the financial partner, and therefore the greater the cost of debt for the company.

The cost of equity (K_E) of a company is not, in reality, a monetary cost, but rather it can be defined as an opportunity cost, that is, a quantity equal to the minimum return required by an investor to finance the company, given the risk conditions that characterise its activity. The greatest difficulties in estimating the cost of equity of a company are therefore linked to the possibility of quantifying the specific risk that weighs on the business activity and of understanding the relationship that exists between this risk and the return expected by the subjects that finance this activity.

The problem has been studied for a long time, and numerous models have been proposed over time in an attempt to identify correct methods of quantifying the cost of equity. However, the issue still represents one of the most delicate and controversial passages, both in the academic and operational worlds, and research is still aimed at determining methods capable of providing a reliable and universally accepted measure.

To date, the most widely used model for estimating the relationship between risk and return is the Capital Asset Pricing Model (CAPM), developed in the 1960s by Treynor (1961),[1] Sharpe (1964), Lintner (1965) and Mossin (1966).

The CAPM is a model that, assuming efficient markets and characterised by homogeneous expectations on the part of investors, proposes to determine the expected return on risk capital through the sum of two addends:

- the risk-free rate, that is, the interest rate offered by investments in risk-free assets, such as government bonds;
- the market risk premium, which can be calculated by multiplying the market risk premium by the relative riskiness of the firm expressed by the coefficient β.

Equation (Eq. 4.2) for calculating the cost of equity is as follows:

$$\mathrm{Ke} = \mathrm{r_f} + \beta\left[\mathrm{E}\left(\mathrm{r_m}\right) - \mathrm{r_f}\right] \tag{4.2}$$

Where:

r_f = Risk-free rate,
$E(r_m)$ = Expected market return,
$E(r_m) - r_f$ = Market risk premium,
β = Risk coefficient of the business activity.

According to the CAPM, therefore, the cost of equity increases linearly as a function of the degree of risk expressed by the coefficient β.

The risk-free rate is calculated on the basis of the yield on government bonds, which are by definition risk-free financial assets.

The second component of the expression represents, as already anticipated, the higher return required by the investor in exchange for the additional risk that participation in the equity capital of a company entails compared to investment in a

[1] Jack L. Treynor's article, "Market Value, Time, and Risk", written in 1961, was only published in 2015 (with minor edits) by Craig William French.

risk-free financial asset. This extra return is measured, in the CAPM, as the product of the market risk premium and the coefficient β.

The market risk premium ($r_m - r_f$) is given by the difference between the expected return on the investment in the market portfolio (r_m), i.e. in the equity capital representing the market as a whole, and the risk-free rate (r_f). Its measurement is usually obtained from the historical series of market returns on shares and government bonds. However, this premium represents the higher remuneration required for the generic investment in risk capital and must therefore be adjusted to take into account the specificities of the individual company being analysed. This correlation occurs through the Beta coefficient, which represents the sensitivity, with respect to market trends, of the return on an investment in a specific equity capital. In other words, it represents a measure of the non-eliminable systemic risk, i.e. that part of the risk linked to the general trend of the economy that the investor cannot remove through a process of portfolio diversification. Investing in shares involves two risk profiles:

- systematic risk, otherwise known as market or non-diversifiable risk,
- the specific risk, otherwise known as residual or diversifiable.

The first is linked to the general trend of the economy and cannot be cancelled by the investor.

The second, however, can be eliminated through a diversification process, which is summarised in the well-known rule don't put all your eggs in one basket.

The effectiveness of diversification is based on the fact that the returns of financial assets are not usually perfectly correlated with each other. For this reason, by building a portfolio of financial assets with different returns and levels of risk, it is possible to maximise the return for a given level of risk or minimise the risk for a given return.

Ultimately, Beta (β) measures how the ROE of the company being analysed reacts to events affecting the market as a whole (system changes).

The Beta of the entire market portfolio is, by definition, equal to 1 and, therefore, securities with:

- Beta >1 are overexposed securities as they present a higher systematic risk than the market. They will therefore have a higher cost of capital than that normally required for a generic risk capital investment (represented by the market portfolio).
- Beta <1 indicates a conservative security, as it presents a lower systematic risk than the market and will therefore have a cost of capital between the risk-free rate and the average cost normally required for investing in risk capital.

It is clear, therefore, that based on the CAPM, the expected excess return for the investment in the company's risk capital varies in a manner directly proportional to the Beta.

Therefore, given a certain risk premium for the generic investment in equity capital ($r_m - r_f$), as the riskiness of the company increases, the Beta increases and, therefore, the cost of capital increases, i.e. the return expected by the investor (K_e), compared to that required for a risk-free investment (r_f).

The factors on which a company's Beta depends are three:

- the business sector,
- the operating leverage,
- the financial leverage.

The business sector influences the Beta as the latter measures the business risk concerning a market index, and consequently, the more sensitive the business sector is to general market conditions, the higher the Beta will be.

The level of Operating Leverage (OL), given by the ratio between fixed costs and total costs, influences the Beta, as the higher the OL, the greater the variability of company results and, therefore, all other things being equal, the greater the sensitivity of the return on capital of the specific company to changes in market conditions. Operating leverage indicates the ability of the company to increase its profits more than proportionally as the turnover grows. A high level of leverage, however, implies greater risk, because in the face of a reduction in volumes and, therefore, in revenues, there is no proportional reduction in costs, mainly fixed. The same goes for financial leverage, which can be calculated as the ratio between debt and equity. The higher this value, the greater the vulnerability of the company in the presence of adverse economic phases and, therefore, the higher the Beta.

Based on this last consideration, therefore, it is easy to conclude that the Beta of a company depends not only on the operational risk, but also on the weight of the debt in the capital structure.

It is important to point out that the cost of capital, a crucial element in determining corporate value, is significantly affected by the ESG policies and practices adopted by companies. Recent literature has further explored the link between ESG scores and the cost of capital, suggesting that careful management of environmental, social and governance issues not only improves corporate sustainability but also translates into tangible economic benefits. Recent studies suggest that companies with high ESG scores tend to benefit from reduced costs of capital. This phenomenon is particularly observed in sectors with high environmental and regulatory risk, where good ESG practices are perceived as an effective risk mitigator. Research by Bannier et al. (2022) and Dallocchio et al. (2025) has shown that improving ESG scores can lead to a reduction in WACC, thanks to more favourable financing conditions and a lower perception of risk by investors.

Furthermore, the impact of ESG scores varies significantly across industry sectors. Sectors such as energy and heavy industry, for example, show a stronger correlation between high ESG scores and reduced cost of capital than low-carbon sectors. This highlights how ESG practices can be particularly beneficial in sectors that are more exposed to environmental and compliance risks.

It is worth adding that sustainability and climate change regulations are becoming increasingly stringent, influencing the financial environment in which companies operate. Companies that anticipate and adapt to these regulations through proactive ESG strategies can thus reduce the risk of future sanctions and new regulations, resulting in a lower cost of capital, as illustrated by Schenker (2020).

I would add that investing in ESG initiatives not only reduces the cost of capital in the short term but also contributes to greater financial stability in the long term. This effect is due to the construction of a reputation for reliability and responsibility, which facilitates access to capital on more advantageous terms and stabilises the company's financial flows over time.

In short, the growing attention towards ESG practices is configured not only as a response to an ethical imperative, but also as a strategic choice for companies that intend to optimise their capital cost structure. The adoption of ESG practices therefore proves to be a strategic investment that, in addition to promoting sustainability, improves the financial resilience and competitive positioning of the company in the long term.

In light of what has been argued so far, it appears clear that ESG performance can prove to be a source of mitigation and/or amplification of the risks identified by the CAPM model: specific risk and systematic risk. In other words, ESG performance can prove to be extremely critical and/or strategic for the company, as it is capable of influencing profitability, value, creditworthiness and share price. These considerations are considered even more valid if we start from the assumption that ESG risk does not represent a type of stand-alone risk, but exerts an influence on financial and non-financial risks. Therefore, company managers are invited to follow a holistic approach, integrating ESG risk into the current processes and metrics used to manage individual risk profiles.

In the literature, we find numerous investigations aimed at understanding the relationships between the ESG world and the cost of capital (equity and debt). Some of these, in particular, aim to investigate the impact of ESG scores (or ESG ratings) on the cost of capital.

Table 4.3 highlights, in summary, the main results emerging from the empirical literature. In this, there are: the number of sampled companies and their country of origin; the survey period; the number of observations; the ESG data providers; the regression method; the correlation, positive (+), negative (−) or not significant (ns), with the cost of capital (WACC: equity and/or debt); the pillars.

Table 4.4 lists the main authors who have analysed the relationship between ESG and company value; in this, there are: the number of sampled companies and their country of origin; the survey period; the number of observations; the ESG data providers; the regression method; the correlation, positive (+), negative (−) or not significant (ns), with the company value; the pillars.

Based on the above arguments, it appears evident that an important channel through which ESG influences corporate results is the cost of capital.

Numerous studies show that high ESG scores are associated with a lower cost of both debt capital (K_D) and equity capital (K_E), with a direct effect on reducing the WACC. In other words, more sustainable companies are generally able to finance themselves at more advantageous rates.

A global analysis conducted by MSCI (Morgan Stanley Complex Index) on approximately 4000 companies from 2015 to 2024 highlights a strong historical correlation between high ESG ratings and lower financing costs. On average, companies with the highest ESG scores had funding costs (Debt and Equity) of around

6.8%, compared to 7.9% for companies with lower ESG ratings in the same period. According to MSCI researchers, companies assessed as more resilient to sustainability risks were systematically able to raise capital at lower costs than those that were more vulnerable. These results confirm previous evidence in the literature: for example, El Ghoul et al. (2011) already empirically documented an inverse relationship between ESG ratings and the cost of equity capital. Similarly, other research has found lower credit spreads and better bank financing conditions for companies with high ESG performance (lower risk perceived by the financier). In short, the trust that sustainable companies inspire in financiers and investors translates into a lower risk premium required, with tangible effects on the containment of the cost of debt and equity. Consequently, virtuous companies from an ESG perspective often benefit from a lower WACC, and therefore from higher net present values of investment projects.

Interestingly, less intuitive evidence also emerges: a high ESG rating does not always imply a lower cost of capital. A recent study focused on companies in 17 countries that are components of the Euro Stoxx 600 index, for example, finds that a one standard deviation increase in the ESG rating can be associated with a slight increase (around 15 basis points) in the implicit cost of equity (Priem & Gabellone, 2024). This effect, observed mainly among large caps (e.g. S&P 500 components), is interpreted by taking into account several factors. First, investing in ESG initiatives entails additional costs that investors may require to be compensated by higher expected returns, thus increasing the cost of equity capital. Furthermore, the adoption of ESG practices may expose the company to new risks (e.g. regulatory risks, possible litigation, or fears of greenwashing). Investors, perceiving such potential risks, may require a higher risk premium. Finally, increasing scrutiny in the ESG market, combined with differences in rating methodologies, can generate uncertainty: disagreement between ESG ratings from different providers tends to increase the level of uncertainty for investors, with possible adverse impacts on value and financing costs. Despite these exceptions and nuances, the overall picture remains that of a beneficial impact of ESG on the cost of capital: companies with strong ESG profiles have access to cheaper capital, thanks to the lower perceived risk and the greater trust they inspire in the market. This strengthens the incentive for management to invest in sustainability, as savings on the cost of capital can more than offset the initial outlays to comply with ESG criteria (so-called win-win of financial sustainability).

Following what has been argued, it seems useful to develop some reflections on the ESG & Corporate Value binomial. The integration of ESG factors in corporate management is now considered a strategic lever for the creation of long-term value. First, companies with high ESG performance tend to benefit from tangible advantages in terms of financial performance, which over time are reflected in corporate value. It is generally accepted that environmental, social and governance initiatives contribute to increasing the value of the company, while improving its operational and income performance (Rong & Kim, 2024). This occurs through multiple channels: for example, effective ESG practices can reduce waste and inefficiencies (higher margins), stimulate innovation and quality (products more appreciated by

Table 4.3 Doctrinal guidelines: correlation between ESG scores (or ESG ratings) and the cost of capital

Authors	Sample	Period	Obs.	Data provider	Regression model	Negative (−) relationship with cost of capital?	Pillar
Sharfman and Fernando (2008)	267 companies (S&P 500)	2001	546	Kinder Lydenberg and Domini (KLD) social performance dataset	Hierarchical	(−) K_E and K_D	E
Chen et al. (2009)	559 companies (17 countries)	2001–2002	989	Credit Lyonnais Securities Asia—CLSA	Data Panel	(−) K_E	G
El Ghoul et al. (2011)	2809 companies (USA)	1992–2007	12,915	KLD STATS	Multivariate	(−) K_E and K_D	E S G
Dhaliwal et al. (2014)	1093 companies (31 countries)	1995–2007	5135	CSR reports	Pooled	(−) K_E	E S G
Suto and Takehara (2017)	481 Japanese companies (2007) 525 Japanese companies (2012)	2008–2013	2680	Toyo Keizai Inc.	Data Panel	(−) K_E (ns) K_D	E S G
Matthiesen and Salzmann (2017)	3439 companies (42 countries)	2002–2013	18,973	Thomson Reuters	Multiple	(−) K_E	E S G
Atan et al. (2018)	54 companies (Malaysia)	2010–2013	162	Bloomberg	Multiple Linear. Least Squares Method—OLS	(+) K_E and K_D	E S G
Gupta (2018)	23301 companies (43 countries)	2002–2012	23,301	Asset4 by Thomson Reuters	Data Panel	(−) K_E	E

(continued)

Table 4.3 (continued)

Authors	Sample	Period	Obs.	Data provider	Regression model	Negative (−) relationship with cost of capital?	Pillar
Bellavite Pellegrini et al. (2019)	182 companies (oil and gas industry)	2002–2018	3094	Thomson Reuters	Easton—with fixed effects	(−) K_E	E S G
Johnson (2020)	68 companies (Johannesburg Stock Exchange)	2011–2018	478	Bloomberg	Data Panel	(−) K_D	E S G
Ellili (2020)	30 companies (Abu Dhabi Stock Exchange and Dubai Financial Market)	2010–2019	300	Bloomberg	Multivariate	(−) K_D	E S G
Nguyen (2020)	80 companies (France)	2006–2017	736	Thomson Reuters	Generalized Method of Moments	(−) K_E	E S G
Wong et al. (2021)	62 companies (Malaysia)	2005–2018	663	Bloomberg	Multivariate	(−) K_D	E S G
Gjergji et al. (2021)	87 companies (AIM Italy)	2018	87	Bloomberg and PMI capital	Data Panel	(+) K_E and K_D	E S G
Nazir et al. (2022)	64 companies (technology sector)	2010–2017	512	Thomson Reuters	Random effects	(+) K_E (ns) K_D	E S G

Source: Personal processing

Table 4.4 Doctrinal guidelines: correlation between ESG scores (or ESG ratings) and company value

Authors	Sample	Period	Obs.	Data provider	Regression model	Relationship "+" "–" "n.s."	Pillar
Velte (2017)	Listed companies (German Prime Standard—DAX30, TecDAX, MDAX)	2010–2014	412	Thomson Reuters	Multivariate	(n.s.) Tobin's Q	E S G
Li et al. (2018)	350 listed companies (FTSE UK)	2004–2013	548	Bloomberg Boardex	Multivariate	(+) Tobin's Q	E S G
Atan et al. (2018)	54 companies (Malaysia)	2010–2013	162	Bloomberg	Multiple Linear. Least Squares Method—OLS	(n.s.) Tobin's Q	E S G
Bhaskaran et al. (2020)	4877 companies	2014–2018	235	Thomson Reuters	Two-stage least squares regression	(+) Tobin's Q	E S G
Al Amosh et al. (2022)	124 companies (Levant countries): Jordan, Palestine, Syria, and Lebanon	2012–2019	883	Bloomberg	Data Panel	(n.s.) Tobin's Q Pillar G (+) Tobin's Q Pillars E, S	E S G
Aydoğmuş et al. (2022)	1720 listed companies (Bloomberg database)	2013–2021	14,043	Refinitiv database	Fixed effects	(+) Tobin's Q Pillar S, G (n.s.ns) Tobin's Q Pillar E	E S G
Chen and Xie (2022)	Listed companies (China)	2000–2020	11,382	Bloomberg	Benchmark, with fixed effects	(+) Tobin's Q	E S G

Source: Personal processing

the market), as well as improve the internal climate and employee productivity. Several empirical studies confirm these effects. Recent studies on international samples show that companies with higher ESG scores, on average, record higher profitability and higher market multiples than their less virtuous competitors. For example, an analysis of S&P 500 companies shows that companies with excellent ESG performance achieve higher market valuations than their peers and better financial results in the medium term (Sharfman & Fernando, 2008). Similarly,

Fatemi, Glaum, and Kaiser (2018) find that transparent disclosure of ESG information amplifies the positive effects of sustainable practices on the financial and market value of the company. In other words, not only does doing well (in ESG terms) create value, but effectively communicating it to stakeholders can further increase market trust and capitalisation. A key element through which ESG creates value is the construction of intangible assets such as corporate reputation and relational capital with stakeholders. Companies that show sensitivity towards the environment, attention to employees, ethical behaviours and transparency of governance tend to enjoy greater trust from customers, investors, communities and authorities. This translates into a sustainable competitive advantage: better customer loyalty, greater ability to attract quality talent and investors, as well as a reputational buffer that mitigates the effects of any missteps. Many authors argue that companies that are more skilled in managing relationships with their stakeholders obtain a high reputational profile, which helps them perform better, especially in times of crisis. The reputational capital accumulated thanks to responsible behaviours can protect the value of the company when the context becomes adverse (e.g. sector scandals, external shocks), acting as an implicit insurance. Furthermore, the solid relationship built with employees, suppliers and social partners creates a climate of trust that allows the company to react with greater flexibility and resilience to challenges, avoiding costly conflicts or operational disruptions. All this generates value. Management is therefore encouraged to continuously improve ESG performance because this increases reputational capital, strengthens the bond with stakeholders and, ultimately, creates value for shareholders.

4.5 Valuing the Firm in the ESG Era

Although ESG has become critical and central factors in the (value) analysis activities of companies, asset managers and investors over time, a critical element remains, I would say fundamental (Chen & Xie, 2022): "*The evaluation of ESG performance is formed in a context where there are no precise, standardised and universally recognised rules*". The result is the lack of a consensus on the relationship between environmental performance and the financial performance of companies.

In particular, it is clear how the misalignment between ESG information and the company's financial performance is further amplified in the presence of an ESG report that is disconnected from the financial one, thus giving credence to the prevailing view that declares the non-financial nature of ESG disclosure.

Therefore, companies and professionals who routinely carry out valuation activities struggle to formulate a valid value proposition for ESG investments, as well as encounter evident difficulties in expressing the value of the company regarding the contribution of ESG factors.

To overcome the aforementioned difficulties, a different recognition of ESG information is required: rather than non-financial, ESG information may be labelled as pre-financial.

Indeed, investors, driven by the desire to obtain a more holistic view of companies' material risks, as well as their growth opportunities and expected profit margins, are increasingly and proactively incorporating ESG factors into their analysis and investment decisions.

In terms of impact, it is clear that ESG factors play a critical and strategic role, as they can influence the company's performance and affect its long-term prospects. For example, in the face of increasingly stringent environmental regulations, CO2-intensive companies could experience higher operating costs and reduced access to credit.

It is commonly accepted that companies that achieve the best ESG performance are less risky and perform better from a profit point of view; characteristics that can affect the value of the company. Therefore, the need to integrate ESG factors in the estimate of the value of the company appears evident.

In terms of evaluation, it is clear that a company committed to improving its ESG position is destined to generate value as a result of better financial performance, which can be represented as follows (Cornell & Damodaran, 2020):

- increase in operating profitability,
- increase in cash flow,
- reduction of the discount rate.

From this perspective, ESG factors can no longer be seen as non-financial aspects, but as pre-financial drivers: leading indicators of a company's ability to generate sustainable profits in the future. Institutional investors and valuers are increasingly actively incorporating these variables into their analyses, recognising that they affect growth prospects and long-term profit margins. In short, the creation of corporate value today also involves the creation of social and environmental value: a sustainable company tends to be more solid and profitable, to the benefit of both stakeholders and shareholders.

As for the variables that affect the determination of the value of a company, among the most accredited methods for the economic evaluation of a company is the Discounted Cash Flow (DCF) financial method, which determines the economic value of a company based on the current value of expected cash flows. The method is based on discounted cash flows and is represented by a function with three fundamental variables: the amount of cash flow for the period considered, the terminal value of the company, and the discount rate. According to this method, the value of the company is given by the sum of two main components: the first consists of the current value of the cash flows of the analytical development period of the forecasts, the second consists of the terminal value at time n, when the analytical forecasting period of the flows ends.

Regarding how ESG aspects influence the variables of the calculation of company value, it is a fact that the value of companies tends to attribute an increasingly significant weight to the so-called internally generated intangibles, including brand reputation and customer base, know-how and research, but other aspects, such as the business model and the way of providing information to stakeholders, also have an increasingly significant impact.

In this context, we would like to highlight how, in the application of the most well-known methodology of corporate value, ESG elements generate their impact on each of the variables that make up the function of the DCF method, which is why it is essential to take them into account in the decision-making process at the highest management levels; let's analyse them:

- *Cash flow discount rate (i).* Among the many factors that influence the discount rate, one of the most relevant is the risk premium, i.e. the percentage that rewards the risk that the saver runs in investing in a specific asset. Among the various forms of risk, such as default, market and equity, the so-called ESG risks are variable, with an increasing weight in influencing the company's performance, particularly in the long term.
- *Terminal Value (TV).* Among the various components of Terminal Value, there are: (i) the growth prospects of the reference sector; (ii) the growth prospects of the company and the historical growth already achieved; (iii) the barriers to entry that exist and the possible threat of substitute products/services/technologies; (iv) the competitive intensity of the sector; (v) brand strength. A relevant component is therefore the capacity for innovation, of product or process: it represents a clear signal of how the company can position itself in the market in the years to come, that is, its ability to adapt and/or anticipate trends.
- *Cash Flow (CF).* This variable represents the measure of cash that the company will be able to generate in an explicit forecast horizon and is linked to an analytical forecast of the flows. Compared to the other variables of the function, ESG variables here can have a relatively lesser impact, but it is still appropriate to take them into account in the analytical plan of the flows, especially where it is necessary or reasonable to make investments in pursuit of objectives with a view to sustainable success from which a reduction in the flows themselves can be generated.
- *Years (n).* The forecast time frame conditions the impact of the value of the first component of the formula, i.e. the impact of the sum of the cash flows, compared to the second component of the Terminal Value and, where the company has a strong impact on ESG variables, n is a factor that must be carefully evaluated to attribute reliability to the analytical plans and their temporal development and, therefore, to their impact on each of the factors of the function of calculating the company value.

In a correct process of informing the market with corporate reporting, disclosure must therefore provide stakeholders and investors with that information and qualitative data connected to ESG issues and sustainability issues, to evaluate the expectation, both of short-term dividends (referred to in the first component of the equation), and of the future value of the company (referred to in the second component of the function as long-term value): the integration of ESG elements in corporate strategies can contribute precisely to the long-term value in the perspective of the sustainable success paradigm.

As highlighted in the significant contribution by Bellavite Pellegrini et al. (2020), the business valuation paradigm is undergoing a substantial evolution after a long period of consolidation on traditional approaches. This evolution is oriented towards models capable of responding to the needs of all stakeholders, whether directly or indirectly involved in the business activity.

The shift to a stakeholder-oriented perspective represents a turning point compared to the classical vision based exclusively on entrepreneurial logics of maximising profit, cash flow and dividend distribution. This vision, dominant especially since the 1970s, finds its best-known expression in Friedman's position (2007), according to which the sole responsibility of the company is to generate profit for its shareholders, provided that it complies with the laws in force and according to ethically correct methods. In the following years, however, the need to integrate ethical and cultural considerations into managerial action has progressively asserted itself, directing resources and strategies towards objectives of social and environmental responsibility, beyond mere economic-financial interests.

The academic and professional debate on the concept of CSR has progressively redefined the very notion of corporate value, leading to a progressive overcoming of the approach centred on shareholder value in favour of that of stakeholder value. In this perspective, stakeholders are defined, according to Freeman (1994), as any group or individual that is crucial to the survival of the organisation and that can influence, or be influenced, by the achievement of its objectives. Gibson (2000) further expands this vision, defining stakeholders as those groups or individuals who have a legitimate interest in the success of a project or operational context, with which the organisation maintains relationships or on which it depends. They can promote or hinder the start of an initiative, based on the perception of its impact, and can even determine its interruption if it is not considered consistent with their expectations or interests.

When approaching the concept of sustainability, especially in the context of corporate assessment, it is essential to have uniform and homogeneous information among different companies, so that they are comparable. However, even with measured data, the creation of value attributable to ESG activities is difficult to determine and/or compare between different companies in different periods, also due to the continuous evolution of the context in which they operate.

In financial markets, operators adopt ESG indices to measure the overall sustainability performance of a company, considering that understanding ESG practices combines the relative performances. In particular, the environmental performance of a company identifies its commitment to reducing resource consumption and emissions, which social respect for human rights, quality of employment, product responsibility and community relations, corporate governance, and finally, the rights and responsibilities of managing a company.

In selecting the main drivers of company value, it is necessary to use some tools well known to the doctrine as well as to professional practice. Among the patrimonial, financial, income-based, and mixed valuation methods, the most widely adopted in practice remains the already mentioned DCF method, based on the discounting of projected cash flows, and the use of multiples, calculated as the ratio between market values and accounting figures.

The DCF method requires the evaluator to rely on internal information produced by senior management, primarily contained in the company's forecast plans and relating to its products, technologies, strategic initiatives, and other forward-looking elements. The use of historical and forecast data allows us to arrive at an adequate and sufficiently dynamic picture of the company. In addition, however, it is also necessary to determine the cost of capital that measures the concept of risk and that must be used as a discounting parameter to discount the prospective cash flows at the reference date. Although, therefore, the DCF presents a notable sacrifice insofar as it is necessary to equip oneself with numerous internal and external information, it is certainly one of the most reliable, or at least less approximate, methods.

The use of multiples, on the contrary, places the comparison with other comparable companies at the centre of the determination of the economic value of the capital. The selection of the group of comparable companies is the most delicate process since belonging to the same sector may not be sufficient. Other comparability factors must be taken into consideration, including company size, product portfolio, degree of internationalisation, geographical location in comparable countries, as well as management and governance characteristics. The simplicity in using the methods represents a strong point (ease of use) but also a weak point (staticity).

In emphasising that the DCF method appears, in abstract terms and when the conditions for its applicability are met, the most appropriate, this belief must also contemplate further elements that impact the cost of capital, namely those non-financial variables of the ESG type. This is because the role of the company no longer consists in the simple allocation of capital, in the use of production factors and in the consequent remuneration of the economic capital invested, but must also contemplate the satisfaction of the interests of all stakeholders, even those who are not exclusively interested in economic-financial performance.

Satisfying these categories of stakeholders mainly concerns intangible, reputational and credibility remuneration in carrying out corporate activities according to an ESG sustainability perspective.

From what has been examined, it is clear that the valuation methods indicated above must be updated in consideration of the changes that have occurred. From this perspective, it could be concluded that the valuation of a company's economic capital cannot depend only on market quantities (multiples) and/or values determined by forecast plans updated at the cost of capital (DCF), but also on ESG performances, which can be considered, positively or negatively, as an element to be added to the known valuation parameters.

In conclusion, the integration of good ESG practices is now a priority on the agenda of entrepreneurs, investors, but also evaluators.

If until recently, attention to ESG issues translated into favouring managerial practices oriented not only to the creation of profit, but also to attention to the environment, eco-efficiency, safety and welfare of workers, transparency, ethics and in general the impact of the company on all its stakeholders, today the ESG paradigm represents in all respects a compass capable of discriminating the long-term sustainability, and therefore the intrinsic value of a company being evaluated.

The interest in everything related to non-financial information shows that such non-quantitative information that can be summarised in ESG indices can be translated into a higher score for the intrinsic value of the company. The overall score, therefore, must be broken down into three sector scores for the three main sectors, which in turn are divided into further elements to be considered in the perspective of company valuation, such as, for example (Paolone & Pozzoli, 2021): (E) Biodiversity/land use; Carbon emissions; Climate change risks; Energy use; Raw material procurement; Regulatory/legal risks; Supply chain management; Waste and recycling; Water management; Weather events. (S): Community relations; Controversial business; Customer/product relations; Diversity issues; Employee relations; Health and safety; Human capital management; Human rights; Marketing and research and development manager; Union relations. (G): Accountability; Anti-takeover measures; Structure and size; Corruption; CEO duality; Executive compensation systems; Ownership structure; Shareholder rights; Transparency; Voting procedures.

Companies oriented towards high ESG performance are those that manage their resources according to environmental sustainability criteria, without giving up investment in innovation as a lever for growth; that believe in CSR, while enhancing the attraction and retention of talent; and that adopt strategies aimed at mitigating corporate risks, maintaining an ongoing dialogue with the entire stakeholder ecosystem. In light of these dynamics, the corporate valuer is called upon to consciously recognise and interpret their overall impact when preparing to estimate the economic value of the company. In this new context, corporate valuation is faced with new challenges, made necessary by the growing attention to sustainability behaviours and the consequent affirmation of an integrated valuation approach, capable of combining financial and non-financial indicators.

The time has therefore come to adopt a holistic approach to business valuation, capable of capturing both the sources of competitive advantage and their evolution over time. This approach is based on the assumption that there is no longer a single dominant driver to which to anchor the fundamental analysis, but rather a plurality of heterogeneous factors, each linked to specific business models and their evolutionary trajectory (Paolone & Pozzoli, 2021).

From this perspective, it seems consistent to believe that, for valuation purposes, the appraisal model can include an ESG premium, to be applied to increase the result obtained through the use of a traditional valuation model, if the company has high environmental, social and governance performance. On the contrary, in the presence of poor ESG performance, this adjustment could be configured in a worsening sense, reducing the estimated intrinsic value.

In any case, accurate and timely communication of non-financial information, particularly related to ESG performance, is crucial to offer investors and credit operators a more comprehensive and transparent representation of the company, from a perspective of mutual benefit (win-win) and valorisation of the distinctive capabilities of the organisation.

The purposes estimated by the assessment of the economic value of the company are therefore strictly connected to the topic in question, requiring a specific reflection on the influence that ESG performances have on the economic capital of the company. From this perspective, the intrinsic value determined through a conventional assessment model can be appropriately adjusted upwards or downwards, on the basis of an analytical examination of the ESG elements that characterise the entity being analysed.

In conclusion, the discipline of business valuation is currently in a transition phase, in which environmental, social and governance factors assume a relevance equal to that of balance sheet indicators. This leads to the urgent need to identify shared best practices, capable of ensuring reliable valuations of high methodological quality and aimed at clearly outlining the near future of the valuation discipline.

As already highlighted by Quintiliani (2021), measuring enterprise value in times of crisis requires a critical reconsideration of traditional valuation approaches, taking into account extraordinary variables that impact operating performance and risk perception. During the pandemic, companies had to deal with unforeseen scenarios, characterised by strong discontinuities in cash flows, regulatory uncertainty, supply chain realignment and significant variations in risk premiums. All this has called into question the assumptions underlying conventional valuation models. In this context, the integration of ESG criteria in corporate valuation methods—in particular in DCF, WACC and Economic Value Added (EVA)—represents a coherent response to the need for stability and reliability. ESG criteria, if properly incorporated, not only positively impact the quality and sustainability of cash flows but also act as levers to reduce the uncertainty perceived by investors and creditors, with potentially favourable effects on the cost of capital. The ESG variable, far from being an external addition to the models, can thus transform itself into a structural factor of evaluation resilience, capable of intercepting weak signals coming from the context and contributing to the construction of more realistic and sustainable scenarios. In short, as highlighted by Quintiliani (2021): "*It is not a question of denying traditional models, but of recognising that value creation today occurs in a more complex context, where economic, environmental and social drivers are increasingly integrated*".

4.6 ESG and Capital Structure: Towards a Sustainability-Aligned Leverage

The relationship between capital structure and ESG factors is a topic of growing relevance both in the academic field and for managers and investors. Capital structure decisions must necessarily incorporate the implications arising from ESG risks and opportunities expressed by stakeholders and financial institutions (Dallocchio et al., 2025).

Classical finance literature, mainly based on dynamic trade-off theory, argues that firms determine their optimal level of leverage by balancing costs and benefits associated with debt and equity (Kraus & Litzenberger, 1973; Myers & Majluf, 1984). Recently, this paradigm has been extended to include ESG variables that, by directly influencing investors' risk perception, alter the optimal corporate capital structure (Ho et al., 2021; Adeneye et al., 2023).

The research by Dallocchio et al. (2025) is directly linked to the classical paradigm of the aforementioned dynamic trade-off theory, integrating and expanding it with the dimension of ESG risks, in particular, climate risks. Specifically, while traditional literature (Kraus & Litzenberger, 1973; Myers & Majluf, 1984) argues that firms pursue an optimal level of leverage by balancing the costs and benefits of debt and equity, the recent integration of ESG variables (Ho et al., 2021; Adeneye et al., 2023) highlights how ESG factors can significantly alter this balance. In this context, the research by Dallocchio et al. (2025) represents a specific advance, empirically demonstrating how the perception and management of risks and opportunities arising from the climate transition directly influence the dynamic behaviour of firms in their search for an optimal capital structure. Indeed, according to Dallocchio et al. (2025), companies that are more engaged in the proactive management of climate risks—and that actively seize the opportunities arising from the ecological transition—are able to reach the desired level of financial leverage more quickly. On the contrary, companies that are less prepared or more vulnerable to regulatory risks are slower in this adjustment process. In short, Dallocchio's study enriches the classic paradigm of the dynamic trade-off theory, clearly showing how the ability of companies to strategically address climate and ESG risks significantly influences their optimal capital structure and the speed with which they are able to reach it.

For greater clarity, companies that adopt proactive strategies towards environmental challenges (climate opportunities) show a significant reduction in leverage adjustment costs and easier access to capital markets (Dallocchio et al., 2025; Ho et al., 2021). This derives from the lower perception of climate risk by investors, which leads to more advantageous financial conditions, highlighting a clear link between high ESG performance and lower cost of capital (Caragnano et al., 2020).

On the other hand, firms that are more exposed to regulatory risks associated with the climate transition face greater difficulties in adjusting their capital structure, due to the perception of a higher default risk (Ginglinger & Moreau, 2023; Li & Zhang, 2023). As a result, these firms face a higher cost of capital and a slower speed of adaptation, which negatively impacts their leverage and overall performance.

These empirical findings underline the need for managers to actively integrate ESG criteria into financial and operational strategies. A passive regulatory approach is not enough: it is necessary to adopt proactive and anticipatory strategies that allow for to exploitation of the opportunities offered by the transition to a sustainable economy, while simultaneously improving the management of financial leverage (Dzieliński et al., 2024).

Effective governance emerges as a crucial element in integrating ESG into corporate financial decisions. Robust governance promotes stakeholder communication, reducing information asymmetries, and improving leverage management through more accurate ESG risk assessment (Zhou & Wu, 2023).

Recent literature highlights a significant differentiation between industrial sectors in terms of exposure to climate risks and the resulting speed of adjustment of the capital structure. Environmentally intensive sectors, such as manufacturing, mining, and utilities, require more rigorous management of financial leverage to mitigate the risks associated with the climate transition and maximise ESG opportunities (Nguyen & Phan, 2020; Bolton & Kacperczyk, 2023).

Regulatory authorities play a crucial role in facilitating access to capital for ESG-friendly companies. Clear and stable regulatory policies help reduce uncertainty and leverage adjustment costs, thus incentivising a faster and more effective transition to sustainable economic models (Sautner et al., 2023).

The combination of capital structure and ESG is an essential pillar in modern corporate management. Companies that effectively anticipate and integrate ESG factors into their financial strategy enjoy a significant competitive advantage while achieving a more balanced and resilient capital structure in the long term.

Chapter 5
Capital Structure, Risk Management and ESG Strategies

5.1 ESG in Strategic Financial Decisions: Capital Structure Reconfiguration

Strategic decisions regarding the capital structure of companies have historically been mainly oriented towards optimising corporate value through the balance between equity and debt capital. However, the increasingly relevant introduction of ESG criteria has significantly changed the traditional paradigm of such decisions, pushing companies to reconsider their financial and operational models from a long-term and sustainability perspective.

In recent years, numerous studies have highlighted how the growing integration of ESG criteria into corporate decisions influences not only the external perception of companies but also their actual ability to create economic value and financial resilience (Bhandari et al., 2022; Miralles-Quirós et al., 2023).

One of the most interesting findings emerging from recent literature concerns the strategic role of the equity component in the capital structure for companies most exposed to ESG risks. Companies operating in sectors highly sensitive to environmental and social issues, such as energy, manufacturing, and resource extraction, must necessarily consider a higher proportion of equity capital compared to debt to reduce their exposure to financial risk.

As highlighted by Lopez-de-Silanes et al. (2024), a capital structure that favours equity allows for mitigating the risks deriving from negative events linked to environmental, social, or governance issues.

At the same time, managing ESG criteria in the capital structure also implies a strategic reassessment of financing sources, favouring innovative financial instruments, such as green bonds, social bonds, and sustainability-linked bonds. The choice to adopt such financial instruments is not simply driven by the need to respond to stakeholder or regulatory demands, but represents a conscious strategy to

A. Quintiliani, *ESG and Value Creation from a Corporate Finance Perspective*, Sustainable Finance, https://doi.org/10.1007/978-3-032-03147-1_5

reduce the cost of capital and improve the competitive positioning of the company on the international financial market (Naeem et al., 2023).

Recently, there has been a strong growth in institutional investors' interest in companies that demonstrate a concrete integration of ESG principles in their capital structure. This phenomenon has led to a significant change in capital flows, increasingly oriented towards companies capable of demonstrating transparency and sustainable performance.

An empirical study conducted by Ginglinger and Moreau (2023) on ESG-driven mergers and acquisitions has highlighted that such operations, in addition to recording superior financial performance, also enjoy greater shareholder stability and lower volatility in returns, precisely thanks to the reduced perception of the associated risk. Furthermore, recent literature highlights how companies with high ESG performance tend to benefit from more favourable financing conditions, including lower interest rates and better contractual conditions. This benefit directly reflects on the company's ability to sustain long-term strategic investments and on the possibility of responding effectively to any financial or environmental shocks (Bolton & Kacperczyk, 2023).

The ESG-oriented financial strategy also affects decisions regarding dividend distribution and payout policy. Companies with a strong ESG orientation generally show a more cautious propensity in distributing dividends, favouring the internal reinvestment of the resources generated in order to finance sustainable projects. This prudent and forward-looking approach tends to further improve the financial resilience of companies, protecting them in times of economic uncertainty (Nguyen, 2020).

It is essential to note that ESG-oriented financial strategies also have significant implications in terms of corporate governance. In particular, greater attention to governance criteria allows for more effective management of financial risk and a better alignment of interests between shareholders, management, and other corporate stakeholders. Recent studies (Sautner et al., 2023) show that companies with solid and ESG-oriented governance structures are able to manage financial and reputational crises more effectively, obtaining a more positive evaluation from investors and markets. Finally, the integration of ESG criteria in strategic capital structure decisions requires companies to review their investment policies. The ability to select and finance sustainable projects becomes a key element to maintain competitiveness in the global market. Companies that demonstrate leadership in the ESG field are able to attract not only greater financial investments but also talent and strategic resources, thus consolidating their competitive advantage in the long term (Zhou & Wu, 2023).

In conclusion, the growing importance of ESG criteria in corporate capital structure does not represent a simple management evolution, but rather a real strategic transformation.

Companies that are able to correctly interpret this change and effectively integrate it into their financial decisions will be able to ensure a lasting competitive positioning, better economic performance, and greater resilience to future economic and environmental challenges.

5.2 Integrated Risk Management: Aligning ESG and Financial Risks

The growing awareness of the financial materiality of ESG risks has led, in recent years, to a profound evolution in the approach to corporate risk management. ESG factors are no longer considered marginal external or reputational aspects, but rather structural components of the overall risk profile of the company. In this context, the integrated management of ESG and financial risks is an essential practice for companies oriented towards strategic sustainability.

One of the most significant developments in recent literature concerns the progressive assimilation of ESG risk to traditional financial risk.

According to Bhandari et al. (2022), the integration of ESG factors in a risk-based decision-making perspective strengthens the resilience of companies to unconventional adverse events, such as climate shocks, environmental disasters, reputational crises, or socio-political tensions. Such events, once considered exogenous, are now integrated into risk management frameworks, also due to the pressure exerted by institutional investors and regulatory authorities. In particular, the study by Chang (2024) shows how the stocks of companies with high ESG ratings are characterised by lower volatility in financial markets, indicating a lower exposure to systemic risk. The implications are clear: ESG practices improve not only operational sustainability but also the stability of financial returns, with positive impacts on the cost of capital.

ESG risk measurement is progressively being structured through quantitative approaches and integrated predictive models. As observed by Gianfrate et al. (2024), there is a growing convergence between ESG models and advanced risk management techniques, including:

- Multiple regression models with ESG variables for the estimation of ESG-adjusted beta.
- Application of modified VaR (Value at Risk) and CVaR (Conditional Value at Risk) for climate scenarios.
- Synthetic indicators derived from machine learning for the identification of non-linear correlations between ESG risk and PD (Probability of Default).

A relevant contribution in this sense is provided by Miralles-Quirós et al. (2023), who analyse how the integration of the ESG score in the credit risk assessment significantly improves the prediction of the Probability of Default, especially in energy-intensive sectors.

Empirical evidence suggests that exposure to unmanaged ESG risks translates into greater financial vulnerability. The study by Albuquerque et al. (2020) documents how firms with poor ESG performance experienced more severe drawdowns during the acute phases of the Covid-19 pandemic crisis. In particular, the S and G dimensions played a key protective role, thanks to their function of safeguarding human capital and governance mechanisms. A similar analysis is conducted by Avramov et al. (2022) and Naseer et al. (2024), who correlate environmental sustainability policies with the probability of experiencing catastrophic events linked to climate risk. Firms more exposed to environmental liabilities tend to record higher risk premiums, with negative effects on fundamental balance sheet metrics.

The adoption of ESG criteria in Enterprise Risk Management systems represents today an advanced frontier of corporate governance. As illustrated by Dobrick et al. (2025), integrated ERM allows companies to map, classify, and mitigate ESG risks as well as operational, market, and credit risks. This integration is facilitated by the adoption of international standards such as the Task Force on Climate-related Financial Disclosures (TCFD) and the updated COSO-ERM framework.

Best practices identified in the most recent literature suggest ESG integration according to a top-down logic, where the board of directors plays a leading role in defining the ESG risk appetite and in supervising integrated sustainability plans (Gillan et al., 2021).

One of the most significant implications of the integration of ESG risks is their direct impact on the formation of prices in financial markets. Bolton and Kacperczyk (2023) demonstrate that investors incorporate an environmental risk premium into the expected returns of bonds, with an increasing differentiation between carbon-intensive and low-carbon companies. This premium, quantifiable through the carbon premium, represents an implicit penalty for companies that do not adopt decarbonisation strategies. In parallel, Li and Zhang (2023) highlight that companies that issue sustainability-linked bonds manage to obtain significantly more favourable financing conditions, thanks to a reduction in the perceived risk spread. These findings confirm that ESG risk is now fully priced by the markets, acting as a reputational and financial lever at the same time.

Financial supervisory authorities are formally recognising ESG risks as potential sources of systemic instability. The European Central Bank (ECB, 2022) and the European Banking Authority (EBA, 2025) have published binding guidelines for the integration of ESG factors into internal credit risk assessment and stress testing models. In particular, climate risk is now recognised as an autonomous driver of financial risk.

A study by Mutarindwa et al. (2024) proposes the introduction of a carbon-adjusted leverage ratio, suggesting that companies should modulate their capital structure according to their carbon intensity to minimise exposure to future environmental regulations and carbon pricing.

Despite the progress, significant critical issues remain. One of the main ones concerns the methodological heterogeneity in ESG ratings. Christensen et al. (2022) show how the correlation between ESG ratings provided by different agencies is often lower than 0.5, making consistent benchmarking between companies difficult. Such ambiguity can generate distortions in risk assessment and hinder effective ESG integration in decision-making processes. Furthermore, excessive emphasis on self-reported metrics risks fostering greenwashing phenomena, compromising the reliability of ESG signals. As highlighted by Berg et al. (2022), data transparency and methodological consistency are essential conditions for effective ESG risk management.

In the context of the integrated management of ESG and financial risks, the pandemic experience is a concrete and instructive point of reference. Covid-19 has highlighted how sudden exogenous events can generate profound and transversal effects on all dimensions of a company, putting into crisis not only operations, but also economic sustainability and long-term resilience. It was an extreme risk, which took on both environmental and social connotations, with evident financial implications. As highlighted by Quintiliani (2021): "*The vulnerabilities of the business system have emerged clearly, and can be summarised in at least three main points: (a) dependence on global supply chains, which have suddenly stopped causing delays or production stops; (b) exposure to financial stress, with consequent difficulty in managing liquidity and cash flows; (c) operational vulnerability, due to the absence of business continuity plans capable of dealing with events of this magnitude*". This risk taxonomy, born from an analysis of the health crisis, today lends itself to being reinterpreted in an ESG key, as it clearly shows the interdependencies between external factors and internal performance. As the author recalls: "*Covid-19 has proven to be an absolute, pervasive, systemic stress event, which has simultaneously affected supply and demand, altering expectations, suspending supply chains, and inducing distortions in individual and collective behaviours*". In light of this, the integration between risk management and ESG criteria requires not only an accurate mapping of potential risks but also a reflection on the systemic and interconnected nature of some events. "*The company is not immune to sudden exogenous events*", Quintiliani (2021) states: "*The latter, if not adequately anticipated, can alter financial and organisational stability, requiring a profound review of decision-making processes*".

5.3 ESG as a Catalyst for Corporate Resilience and Long-Term Value

In recent years, the concept of financial resilience has acquired a central role in the academic and managerial debate, becoming an essential attribute to ensure the operational continuity of companies in contexts characterised by volatility, uncertainty, and exogenous shocks. At the same time, the awareness that ESG practices can act as real drivers of resilience has strengthened, helping to strengthen the ability of companies to adapt, absorb, and react effectively to economic-financial, health, or climate crises.

Traditionally, financial resilience has been defined as the ability of an organisation to maintain economic and financial equilibrium even in the presence of shocks, thanks to solid capital indicators, available liquidity, and the ability to access credit (Barbera et al., 2021). However, the most recent literature proposes a broader and more systemic vision, in which financial resilience also includes intangible elements such as stakeholder trust, reputational capital, management quality, and long-term strategic orientation (Duchek et al., 2020). In this perspective, the integration of ESG criteria is configured not only as a sustainability measure but also as a strategic lever to strengthen corporate financial resilience. The central thesis supported by recent studies (Fatemi et al., 2018; De Giuli et al., 2024) is that companies with high ESG quality are less vulnerable to market shocks and more capable of maintaining stable cash flows, margins, and self-financing capacity.

Several empirical studies confirm the positive association between ESG performance and financial resilience during periods of economic turbulence. Albuquerque et al. (2020), analysing data from US companies during the first wave of the Covid-19 pandemic, observe that companies with better ESG scores experienced lower stock losses, lower volatility, and a faster recovery in market values. The authors interpret this result as a reflection of greater investor confidence in companies perceived as more responsible, solid, and long-term oriented.

Similar analyses have been conducted by Lins et al. (Lins et al., 2017), who highlight how social capital intensity—that is, the intensity of the company's social commitment—is positively correlated with operational stability in times of crisis. Companies that invested in human capital, employee relations, and local communities before the pandemic crisis showed greater internal cohesion and a more effective capacity to respond to the emergency.

A cross-country study by Gianfrate et al. (2024) extends the analysis to the international level, showing that the protective effect of ESG practices is more marked in markets characterised by lower institutional protection and greater regulatory instability. This suggests that, in fragile contexts, the voluntary adoption of ESG standards can represent a form of self-insurance for the firm, reducing idiosyncratic and systemic risk.

Financial resilience cannot be separated from the resilience of the business model. In this perspective, ESG practices strengthen the sustainability of the business model, contributing to improving transparency, the quality of governance, and the solidity of relationships with stakeholders. According to Eccles et al. (2022), companies that have integrated ESG strategies into their operations show less exposure to negative reputational events, lower compliance costs, and greater adaptability to regulatory changes.

Another relevant element is strategic coherence. As documented by Krueger et al. (2020), companies that implement ESG policies that are consistent with their core business are rewarded by the market in terms of stability and resilience. On the contrary, behaviours perceived as opportunistic or misaligned (greenwashing) determine greater volatility of returns and an increase in reputational risk.

Financial resilience is also manifested in the ability to maintain stable and predictable cash flows. Several studies show that companies with high ESG ratings have lower earnings variability, greater ability to generate operating cash, and less dependence on debt capital in adverse market conditions (Rhou et al., 2016; Broadstock et al., 2021). These results are partly explained by better management of operational risks, greater energy efficiency, and a more flexible cost structure, typical characteristics of sustainable companies. In particular, the study by Broadstock et al. (2021) found that during the Covid-19 pandemic, companies with high ESG scores showed superior performance in financial markets, thanks to their greater ability to manage systemic risk. Similar results emerged from the study by Gianfrate et al. (2024), according to which portfolios composed of companies with high ESG ratings experienced lower drawdowns in phases of market stress, consolidating the role of ESG as a key element for the protection of corporate value.

One of the most systematic analyses on the topic is provided by Wang et al. (2024), who analysed a sample of over 4500 Chinese companies, demonstrating that ESG performance strengthens corporate resilience through three main channels: reducing market value volatility, strengthening corporate reputation, and improving stakeholder relations. These results are consistent with the growing emphasis placed in the literature on ESG as a preventive risk management mechanism and as a lever for building solid social capital. Confirming the strategic relevance of ESG in terms of resilience, the study conducted by Dalal and Thaker (2019) highlights how the integration of sustainable practices significantly reduces the cost of capital, improving access to credit and financial stability, especially in periods of climate crisis or macroeconomic turbulence. This effect is attributed to the perception, by investors and creditors, of a lower riskiness of companies with solid ESG policies.

A particularly innovative line of research is represented by the assessment of ESG engagement in European financial markets during periods of instability. According to the study by Ricci et al. (2024), ESG stocks have shown a more stable behaviour than traditional indices during exogenous shocks, positioning themselves as safe havens for investors, especially in more advanced regulatory contexts. In operational terms, the adoption of ESG practices allows companies to more effectively address value chain disruptions, mitigate reputational and regulatory risks, as well as adapt more quickly to regulatory and market changes. ESG-oriented companies demonstrate a greater ability to preserve margins and liquidity, thus improving their financial solidity in times of greater vulnerability. Further evidence emerging from recent literature concerns sectoral resilience. Deng's study (2022) confirms that the protective effect of ESG is more marked in capital-intensive and highly environmentally exposed sectors, such as energy, manufacturing and chemicals, where ESG risk management is crucial for competitive survival. Finally, according to the systematic review by Wahyuni et al. (2024), ESG integration is now considered a fundamental dimension of financial risk management. It not only allows for better management of operational and reputational risks but also promotes alignment with regulatory and market expectations in a context increasingly oriented towards sustainability and integrated disclosure.

A study conducted on a sample of European companies by El Ghoul et al. (2011) reveals that free cash flow volatility is inversely related to the aggregate ESG score, with a particularly significant effect for the E and G dimensions. Environmentally responsible companies tend to be less exposed to sanctions, supply chain disruptions or unpredictable increases in production costs.

One of the intangible assets that contributes most to financial resilience is reputational capital. The most recent literature highlights that companies with strong ESG credentials enjoy greater trust from customers, suppliers, workers, and financiers (Hoepner et al., 2024). This trust translates into more stable relationships, better contractual conditions, and greater tolerance in phases of temporary underperformance.

ESG reputational capital, therefore, acts as a buffer that absorbs external shocks. This protective function has also been documented in the case of industrial scandals: studies such as Liang and Renneboog (2020) show that, in the event of negative events, companies with high ESG scores experience less pronounced reductions in market prices than less virtuous companies.

It should also be emphasised that the effect of ESG practices on financial resilience is not static but varies according to the life cycle of the company. The evidence of Wang et al. (2023) indicates that companies in the early stages of development benefit more from ESG policies in terms of attracting patient capital and reducing the risk of failure. On the contrary, in mature companies, the effect manifests itself mainly in greater efficiency in cost management and stability of returns. This dynamic perspective requires particular attention to the temporal coherence of ESG policies. Resilient companies are those that adopt an integrated and proactive vision of sustainability, capable of evolving with the needs of the market and stakeholders (Lin et al., 2025).

In summary, the most recent academic literature provides solid evidence of the crucial role played by ESG criteria in strengthening corporate financial resilience. Companies that consistently integrate ESG practices are better able to face adverse events, reduce exposure to systemic risk, and maintain more stable financial performance over time. These benefits derive from a plurality of factors: better operational management, consolidated reputation, stakeholder trust, and more flexible and sustainable business models. In a context marked by recurring global crises—health, energy, climate, and geopolitical—the ability to build and maintain financial resilience through ESG integration represents not only a good practice, but a strategic necessity for the survival and prosperity of companies in the long term.

A further element that strengthens the role of ESG in building corporate resilience is represented by the lessons learned during the Covid-19 pandemic crisis. In this context, the business system had to deal with a sudden global exogenous shock, which called into question both traditional management models and financial planning practices. Experience has shown how the use of scenario planning and predictive analysis tools is a necessary condition for competitive survival, especially in conditions of systemic instability. It is no coincidence, as observed in a previous contribution by Quintiliani (2021): "*Scenarios are useful for understanding what may happen. In this way, the company will be able to evaluate, through the planning tool, where it is today about the scenario*". Added to this is the relevance of the temporal scan, understood as an analytical practice capable of systematising threats, opportunities, and future implications within an integrated business continuity logic. Therefore, as stated by Quintiliani (2021): "*The company will have to equip itself with a system for planning complex emergencies, such as the one triggered by Covid-19, taking care of the temporal scanning of the processes and the related evaluation phases*". In this framework, the role of the Chief Financial Officer (CFO) emerges forcefully as a strategic player in the management of financial needs, in the monitoring of treasury tensions, and in the recalibration of financing sources according to short and medium-term equilibrium logics. As highlighted, in fact, by Quintiliani (2021): "*The CFO is called to play a central role in crisis management, carefully planning financial needs, monitoring treasury tensions and remodulating financial procurement strategies*". ESG-oriented financial resilience is therefore nourished not only by ex post indicators, but above all by ex-ante systemic capabilities, based on risk culture, aware governance, and adaptive models. In other words: "*Behavioural models capable of strengthening the company's immune system [...], capable of reaching a state of grace, in perfect harmony with the world around it*". These models presuppose a company management that can distinguish itself, as stated by Quintiliani (2021), for: "*Foresight, ability to recognise valuable information, reactivity and ability to modify processes*". This evidence, drawn from one of the most recent global crises, confirms the need to place the integrated management of ESG risks within a dynamic vision of corporate resilience, capable of combining ethics, strategy, and financial solidity.

Furthermore, Quintiliani (2021) continues: "*Today, post Covid-19, business conditions are changing rapidly. Therefore, the adoption of flexible strategies is considered desirable; in other words, strategies that allow companies to learn and adapt their forms and expectations to each growth opportunity*". The integration of ESG factors into the work of company management can act as a catalyst for this adaptive transformation, to the extent that it promotes a long-term vision, greater sensitivity to systemic risk, and a managerial culture oriented towards sustainability. However, this evolutionary capacity is not automatic and presupposes the overcoming of deeply rooted cognitive barriers. As Quintiliani (2021) recalls: "*The entrepreneur will have to [...] shake off any prejudice that legitimises the irrational maintenance of the status quo. Human beings will have to be able to imagine the impossible*". It follows that company management, to be truly effective in terms of ESG, must be configured as a cognitive platform capable of recognising and managing complexity. This implies the development of skills oriented towards foresight, proactive reading of the context, and conscious modification of business processes. As already highlighted, resilient companies are those capable of relying on (Quintiliani, 2021): "*Foresight, ability to recognise valuable information, innovative capacity and reactivity in the face of a changing environment*".

5.4 ESG Ratings and Their Effects on Financial Cost Formation

The growing interest in sustainability has placed the link between ESG ratings and the cost of capital, i.e., the effect that the evaluation of ESG performance has on the ability of the company to finance itself at advantageous conditions, at the centre of academic and professional debate. This relationship is crucial not only for internal financial management but also for understanding how markets incorporate ESG factors in determining risk and in the formation of the price of capital, both debt and equity.

ESG ratings, provided by specialised agencies such as MSCI, Sustainalytics, Refinitiv, ISS, and Moody's ESG Solutions, represent a synthetic measure of the exposure and management capacity of companies to non-financial risks. Despite the methodological heterogeneity between rating providers, they are increasingly used by institutional investors, banks, and insurance companies as determining variables in risk assessment models, in particular for credit analysis and estimation of the cost of equity capital.

As Christensen et al. (2022) point out, ESG ratings should not be understood as simple reputational tools, but as increasingly integrated components in risk pricing systems, as they reflect the company's ability to sustain its operating and financial model over time in the presence of factors of climatic, social or regulatory instability.

Numerous studies document a negative association between ESG ratings and the cost of corporate debt. Companies with high ESG ratings are more attractive to financial institutions, which perceive them as less exposed to operational, legal, and reputational risks and therefore deserving of more favourable credit conditions.

Drobetz et al. (2024) show that a single point improvement in a company's ESG rating is correlated with a statistically significant reduction in bond spreads. This effect is particularly evident in corporate bonds issued in European markets, where regulation and ESG sensitivity are higher.

Similarly, other studies (Cheng et al., 2023; Goss & Roberts, 2011) observe that carbon-intensive firms obtain an immediate benefit in terms of reduced financing costs when they improve their ESG disclosure and environmental rating. These results reinforce the thesis according to which sustainability acts as a risk mitigator for creditors, improving perceived creditworthiness.

The cost of equity is influenced by investors' perceptions of the future risk of the company. In this context, the ESG rating acts as a signal of management quality and prospective solidity. Several studies show that companies with high ESG ratings have lower volatility in expected cash flows and higher long-term shareholder loyalty.

In particular, some studies (Apergis et al., 2022; Gonçalves et al., 2022) analyse large samples of international companies and confirm that ESG ratings are negatively associated with the cost of equity, especially in highly regulated sectors. These studies clarify that ESG information reduces information asymmetry and attenuates uncertainty perceived by investors, thus contributing to the reduction of the required risk premium.

The study by Avramov et al. (2022), instead, introduces the concept of uncertainty-adjusted ESG premium: when ESG ratings are consistent across agencies, the cost of capital decreases more markedly; on the contrary, divergence between ratings attenuates or cancels the positive effect. This result underlines the importance of transparency and methodological harmonisation in the ESG ratings market.

The spread of sustainability-linked bond instruments—such as green bonds, social bonds, and sustainability-linked bonds—has provided an important empirical basis for assessing the effect of ESG ratings on the returns required by investors (Berdiev, 2025).

Bolton and Kacperczyk (2023) show that green bond issuers with high ESG ratings are able to place their issues at significantly lower rates than their less sustainable counterparts, even with the same traditional credit rating. This result indicates that the bond market incorporates ESG ratings in a complementary but independent way compared to traditional risk metrics. Furthermore, recent studies such as Li and Zhang (2023) confirm that the demand for ESG bonds is higher and more stable, especially in phases of financial turbulence, helping to contain the volatility of financing costs over time.

The interaction between ESG ratings and the cost of capital is increasingly mediated by regulatory interventions that push financial intermediaries to incorporate sustainability factors into credit allocation processes and risk management. The SFDR and the CSRD of the European Union, for example, oblige funds and companies to provide evidence of the ESG profile of their investments and operations.

As reported by Gianfrate et al. (2024), regulation contributes to strengthening the signalling role of ESG ratings, making them an integral part of credit scoring models and due diligence processes. This implies a growing alignment between declared sustainability and the economic conditions of access to capital.

Despite the growing impact of ESG ratings on the cost of capital, the literature has raised concerns about the poor convergence of methodologies adopted by rating agencies. Berg et al. (2022) find a correlation lower than 0.5 among the main ESG rating providers, signalling the risk of information arbitrage and poor comparability between ratings. This issue is particularly relevant for investors who base asset allocation decisions on ESG indicators, as low reliability of metrics can lead to pricing errors, distorted valuations, and allocative inefficiencies. Some studies propose the adoption of composite ratings or weighted averages between different sources, but the issue remains open in the absence of regulatory standardisation.

From a management perspective, the link between ESG ratings and the cost of capital means that companies can no longer consider sustainability performance as an ancillary element. On the contrary, they must be conceived as an integral part of the financial strategy. Companies that invest in consistent, transparent, and measurable ESG initiatives obtain a tangible competitive advantage in terms of the cost of financing, attractiveness for institutional investors, and stability of the capital raised.

According to Hoepner et al. (2024), management should aim for a proactive governance of ESG disclosure, promoting initiatives that improve the quality and credibility of reported data, in line with financial market expectations.

In summary, the most recent literature shows a growing consensus that ESG ratings significantly influence the cost of capital, both debt and equity. Firms with high ESG ratings tend to benefit from more efficient and less costly access to financing, thanks to lower risk perception and greater investor confidence. However, critical issues related to methodological heterogeneity and the information quality of ratings persist, which require a strengthening of disclosure practices and more stringent regulation. For sustainability-oriented firms, the ESG rating is not just a reputational indicator, but a real determining factor in the capital structure.

Chapter 6
Literature Review (Years: From 2020 to 2025—Journal: Indexed Scopus and WoS)

6.1 ESG and Corporate Profitability: Empirical Evidence

The relationship between ESG indicators and financial performance is a topic of increasing interest in both academic literature and the business world. The growing attention towards sustainability has led companies to integrate ESG practices into their strategies, with the hope of improving not only social and environmental impact but also financial performance. This review explores recent studies from 2020 to 2025 to analyse the empirical evidence and methodologies used.

01.

Title: Exploring the relationship between ESG, financial performance, and corporate reputation using ANOVA: The case of Colombian companies.

Authors: Guillermo D. Palacin-Bossa, Luis G. Alvear-Montoya, Mayra A. Macías-Jiménez.

Published: 2024 (Procedia Computer Science, vol. 241, pp. 552–557).

URL: https://doi.org/10.1016/j.procs.2024.08.079.

Summary: The study analyses the relationship between ESG performance, financial performance, and corporate reputation in 13 large Colombian companies included in the COLCAP index, using analysis of variance (ANOVA), simple and multiple regression methods. The results indicate a strong positive correlation between ESG and corporate reputation, while no significant linear relationships emerge between ESG and financial performance indicators (ROA and ROE). The results suggest that the evaluation of the effectiveness of ESG practices should extend beyond traditional financial metrics, including intangible dimensions such as reputation.

Theoretical framework: The article is explicitly based on the Stakeholder Theory (Freeman, 2010) as a conceptual basis to justify the relevance of corporate reputation as an outcome of ESG practices. According to this theory, value creation is not limited to shareholders but involves all stakeholders, and reputation is considered a

A. Quintiliani, *ESG and Value Creation from a Corporate Finance Perspective*, Sustainable Finance, https://doi.org/10.1007/978-3-032-03147-1_6

form of relational capital that can be enhanced by the adoption of responsible behaviours on the environmental, social, and governance level.

Research Questions & Hypotheses:

- Research Questions. The study investigates the following main questions:

  What is the relationship between ESG and financial performance (ROA and ROE) in Colombian companies?
  What is the relationship between ESG and corporate reputation (REP)?
  Does average reputation across economic sectors influence the ESG-performance relationship?

- Hypotheses. The article explicitly formulates the following main hypotheses:

  H1: ESG has a non-linear relationship with financial performance.
  H2: ESG is positively associated with corporate reputation (REP).

Empirical findings (confirmation or not of the hypotheses):

H1: Partially confirmed. There is no significant linear relationship between ESG and ROA or ROE; the correlation coefficient is weak and negative.
H2: Confirmed. The correlation between ESG and corporate reputation is strongly positive ($r = 0.977$).

Research method: (i) Statistical analysis with ANOVA, MANOVA; (ii) Simple and multiple linear regression; (iii) Pearson correlation analysis.

Period: 2018–2023.

Sample & Data:

- 13 large companies included in the Colombian COLCAP index.
- Industry: Oil & Gas (38%), Financial Services (31%), Retail (8%), Services (8%), Others (15%).
- Sources: certified financial statements and Merco database (ESG and reputation).
- Total observations: 78.

Variables & Controls:

Dependent:

- ROA.
- ROE.

Explanatory:

- ESG score (Merco Responsabilidad ESG).

Controls:

- Sector.
- Time (tested but not significant).

Others:

- Corporate reputation (REP, obtained from Merco Empresas)—treated as a mediating variable.

02.

Title: Impact of ESG performance on firm value and profitability.

Authors: Mahmut Aydoğmuş, Gü zhan Gülay, Korkmaz Ergun.

Published: 2022 (Borsa Istanbul Review, 22(S2), pp. S119–S127).

URL: https://doi.org/10.1016/j.bir.2022.11.006.

Summary: The study analyses the impact of ESG performance on firm value (Tobin's Q) and profitability (ROA), using a large international dataset. The results show that the combined ESG score is positively and significantly associated with both firm value and profitability. However, only the Social and Governance dimensions show significant effects on Tobin's Q, while all ESG pillars (including Environment) positively influence profitability. The paper contributes to the ongoing academic debate on the relationship between sustainability and financial performance, supporting the stakeholder theory.

Theoretical framework: Also, in this case, the theoretical framework of reference is the Stakeholder Theory. The conceptual transition from the shareholder-centric model to a stakeholder-inclusive approach is also recalled, as promoted by the Business Roundtable and in line with the growing relevance of ESG metrics in global financial markets. The article implicitly contrasts this approach with the Shareholder Theory (Friedman, 2007), which is recalled only to explain the contrasting results present in the literature.

Research Questions & Hypotheses:

- Research Questions:

  Is ESG performance positively associated with enterprise value?
  Is ESG performance positively associated with profitability?

- Hypotheses:

  H1: ESG scores have a positive and significant impact on enterprise value (Tobin's Q).
  H2: ESG scores have a positive and significant impact on profitability (ROA).

Empirical findings (confirmation or not of the hypotheses):

H1: Partially confirmed. Combined ESG, Social, and Governance are significantly positive; Environment has no significant effect on Tobin's Q.
H2: Confirmed. All ESG pillars (combined, Environment, Social, Governance) show a positive and significant impact on ROA.

Research Method: (i) Panel data with fixed effects; (ii) Robust regressions with correction for heteroskedasticity and autocorrelation (White robust estimator).

Period: 2013–2021.

Sample & Data:

- 1720 listed companies globally.
- 14,043 firm-year observations.
- Sources: Bloomberg (Tobin's Q, ROA, control variables) and Refinitiv (ESG scores).

Variables & Controls:

Dependent:

- Tobin's Q (enterprise value).
- ROA (Return on Assets, profitability).

Explanatory:

- ESG Combined Score (ESG_CS).
- Environment Score (ENV).
- Social Score (SOC).
- Governance Score (GOV).

Controls:

- Size (Logarithm of Total Assets—Log_TASST).
- Leverage (Total Debt/Total Assets—TDTA).

Others:

- Environmental, social, and governance variables obtained from Refinitiv, updated annually.
- Statistical robustness tests: ADF, Hausman, Breusch-Pagan LM, Wooldridge, Pesaran CD.

03.

Title: ESG–Firm Performance Nexus: Evidence from an Emerging Economy.

Authors: Ajithakumari Vijayappan Nair Biju, Sreelekshmi Geetha, Salu Prasad, Aghila Sasidharan, Ambili Jayachandran.

Published: 2025 (Business Strategy and the Environment, 34(3), pp. 3469–3496).

URL: https://doi.org/10.1002/bse.4152.

Summary: The study explores the relationship between ESG performance and financial performance in Indian firms included in the NIFTY 100 ESG index for the period 2014–2023. The analysis is articulated in two directions: on the one hand, it assesses the impact of ESG (and its components E, S and G) on firm performance; on the other hand, it investigates the market determinants that influence the adoption of ESG practices, with particular attention to the moderating role of governance (board independence). The results highlight the absence of a significant direct relationship between ESG and financial performance, while governance and market forces (e.g., VIX, green bonds) are relevant determinants of ESG performance.

Theoretical framework: The article adopts a multiple theoretical framework, integrating three main approaches: (i) Stakeholder Theory (Freeman, 2010): It is the main foundation used to explain the adoption of ESG practices as a response to the growing pressure from stakeholders. The authors underline that sustainability performance can positively influence the firm's value in the long term precisely because it aligns the interests of the firm with those of all stakeholders, and not only shareholders; (ii) Agency Theory (Jensen & Meckling, 1976): It is invoked to explain how ESG initiatives can generate conflicts of interest between managers and shareholders, leading to potential agency costs, especially when the initiatives are driven by opportunistic or facade purposes (greenwashing); (iii) Signalling Theory: Used to interpret ESG practices (e.g. green bond issuance, ESG disclosure) as signals to the market and investors, intended to improve the perception of transparency and commitment to sustainability. According to this theory, effective ESG communication can reduce information asymmetry between firms and investors, influencing market value judgments. These three theories are harmonised in the work to explore both the impact of ESG on financial performance and the role of market factors and governance (e.g., board composition) as moderators and determinants of ESG performance.

Research Questions & Hypotheses:

- Research Questions:

  Does ESG performance impact the financial performance of emerging market companies?
  Do board characteristics (independence) moderate the ESG-performance relationship?
  Do market forces (volatility, uncertainty, green bonds) influence ESG performance?

- Hypotheses:

  H1: ESG performance positively influences financial performance.
  1a–1c: The same applies to the E, S, and G components.
  H2: Board independence positively moderates the ESG-performance relationship.
  2a–2c: The same applies to E, S, and G.
  H3: Market forces positively influence ESG performance.
  3a–3c: The same applies to the E, S, and G components.

Empirical findings (confirmation or not of the hypotheses):

H1: Not confirmed. No statistically significant relationship between ESG and financial performance (Tobin's Q).
H2: Partially confirmed. Board independence positively moderates the ESG-performance link, except for the Social component.
H3: Confirmed. Market forces significantly influence ESG performance: VIX (volatility): positive effect on ESG; WUI (global uncertainty): negative effect on ESG; Green bonds: positive effect on ESG, S and G (not significant on E); Beta (systemic risk): no significant influence.

Research Method: (i) Panel regression models with fixed effects; (ii) Moderating interactions; (iii) Dumitrescu–Hurlin causality tests, and marginal impact analysis.

Period: 2014–2023.
Sample & Data:

- 87 Indian companies in the NIFTY 100 ESG index.
- 676 firm-year observations.
- Sources: Refinitiv Eikon and Bloomberg.

Variables & Controls:

Dependent:

- Tobin's Q (proxy for firm performance—market value).

Explanatory:

- ESG aggregate score.
- E, S, G individual scores.

Moderator:

- Board independence (percentage of independent members).

Explanatory (second model):

- VIX (India), World Uncertainty Index (WUI), Green Bond Issuance (GBI), Beta (systemic risk).

Controls:

- Company size (market cap).
- Leverage.
- Sales growth.
- Board size.
- Board tenure.

Others:

- ESG is both a dependent (second part) and an independent (first part) variable.
- ESG ratings from Refinitiv.

04.
Title: An empirical analysis of the impact of ESG on financial performance: The moderating role of digital transformation.
Authors: Tao Fu, Jiangjun Li.
Published: 2023 (Frontiers in Environmental Science, 11, Art. 1256052).
URL: https://doi.org/10.3389/fenvs.2023.1256052.
Summary: The study analyses the effect of ESG performance on the financial performance of Chinese A-share listed companies, considering digital transformation as a moderating variable. The results indicate that ESG has a positive and significant impact on profitability (ROA), and that this effect is strengthened by a high degree of corporate digitalisation. Moreover, the effect is heterogeneous: it is stronger in non-state-owned enterprises, in eastern regions, and in those operating in

polluting sectors. The study proposes an integrated ESG–digital approach to promote sustainable corporate development in emerging markets.

Theoretical framework: The article is based on an integrated theoretical framework that combines three main perspectives: (i) Stakeholder Theory (Freeman, 2010): It is the central theoretical reference used to explain the relationship between ESG and financial performance. According to this theory, careful management of stakeholders' interests, including those of employees, suppliers, communities, customers, and public authorities, fosters the creation of sustainable value and enhances financial performance. ESG is conceived as a tool to strengthen corporate legitimacy and reputation; (ii) Resource-Based View (RBV): Implicitly recalled in the analysis of the role of digital transformation. The authors interpret digital capabilities as a strategic resource capable of enhancing the effectiveness of ESG practices. Digital transformation allows for improved operational efficiency, data collection and transparency, favouring better ESG results and, ultimately, better financial performance; (iii) Theory of Information Asymmetry & Signalling Theory: Digital technologies (e.g. big data, blockchain) reduce information asymmetry between companies and stakeholders, facilitating more transparent and reliable communication of ESG performance. The adoption of digital tools for ESG is thus seen as a positive signal to the market, capable of improving investor confidence and corporate competitiveness.

Research Questions & Hypotheses:

- Research Questions:

  Does ESG improve corporate financial performance?
  Does Digital Transformation Drive ESG Adoption?
  Does digital transformation moderate the ESG–financial performance relationship?

- Hypotheses:

  H1: Better ESG performance improves financial performance (ROA).
  H2: Digital transformation improves ESG performance.
  H3: Digital transformation strengthens the positive effect of ESG on financial performance.

Empirical findings (confirmation or not of the hypotheses):

H1: Confirmed. ESG has a positive and significant impact on ROA (coeff. ≈ 0.894, $p < 0.01$).
H2: Confirmed. Digital transformation is positively associated with ESG performance (coeff. ≈ 2.518, $p < 0.01$).
H3: Confirmed. The ESG × Digital Transformation interaction is significant and positive, with a moderating effect on financial performance.

Heterogeneity of results: The ESG → ROA effect is significant only for: (i) Non-state-owned enterprises (ESG coeff. ≈ 0.217, $p < 0.05$); (ii) Eastern regions of China; (iii) Polluting enterprises (ESG coeff. ≈ 0.459, $p < 0.05$).

Research Method: (i) OLS regression on unbalanced panel data; (ii) Interaction test (moderator); (iii) Robustness analysis with alternative ESG variable (Wind ESG_1); (iv) Heterogeneity analysis (ownership, geographical area, polluting sectors).

Period: 2015–2021.

Sample & Data:

- 2256 Chinese listed companies (A-share).
- 15,710 firm-year observations.
- ESG score: Huazheng Rating (scale 1–9).
- Financial data: CSMAR database.
- Digital transformation: textual analysis of annual reports (keyword frequency with Python).

Variables & Controls:

Dependent:

- ROA (Return on Assets).

Explanatory:

- ESG score (1–9).
- ESG_1 (variant for robustness testing).

Moderator:

- Digital Transformation Index (DTB), built from keywords in company reports.

Controls:

- Size (total active logs).
- Debt (log debt/total assets).
- Leverage (log fixed assets/total assets).
- Age (log years of activity).
- Cash (liquidity on assets).
- Balance (shareholder concentration).
- Wage (executive compensation).
- Regional GDP.
- Year and sector fixed effects.

Others:

- Company classification by: (i) Ownership type: state-owned/non-state-owned; (ii) Region: east/central-west; (iii) Sector: polluting/non-polluting.

05.

Title: Does ESG always improve corporate performance? Evidence from a firm life cycle perspective.

Authors: Shang Gao, Fanchen Meng, Wenshuai Wang, Wenxin Chen.

Published: 2023 (Frontiers in Environmental Science, 11, Art. 1105077).
URL: https://doi.org/10.3389/fenvs.2023.1105077.

Summary: The study examines whether and how ESG performance (ESGP) affects corporate performance, distinguishing between different stages of the firm life cycle (growth, maturity, decline). By analysing 11,742 firm-year observations of 1486 Chinese A-share firms (2010–2020), the mediating role of financial risk and the moderating effect of three variables: disclosure quality, financial flexibility, and environmental uncertainty are analysed. The results confirm the positive effect of ESG on performance, especially in the growth phase, and the mediation of financial risk only in the growth and maturity phases.

Theoretical framework: The article adopts an integrated theoretical approach mainly focused on three conceptual pillars: (i) Firm Life Cycle Theory (Miller & Friesen, 1984; Adizes, 1990; Dickinson, 2011). It constitutes the main theoretical foundation of the study. Firms are considered living organisms that go through distinct phases—growth, maturity, and decline—with specific implications for ESG strategies. The authors analyse how the effectiveness of ESG varies according to the firm's life cycle phase, highlighting that the costs/opportunities of ESG practices are heterogeneous based on the firm's evolutionary stage; (ii) Signalling Theory: ESG and non-financial disclosure are interpreted as positive signals to the market. Proactive communication of ESG commitment reduces information asymmetry and improves corporate reputation, influencing investors' trust; (iii) Financial Risk Mitigation Framework: ESG is also seen as a financial risk mitigation tool. In particular, the study examines the mediating role of financial risk (measured by Altman Z-score) in the ESG-performance relationship. The model hypothesis is that better ESG contributes to reducing risk and, consequently, to improving financial performance, especially in the growth and maturity phases.

Research Questions & Hypotheses:

- Research Questions:

  Does ESG improve corporate performance across all lifecycle stages?
  Does financial risk mediate the relationship between ESG and corporate performance?
  Do internal/external conditions (disclosure, financial slack, uncertainty) moderate this relationship?

- Hypotheses:

  H1: ESGP is negatively related to corporate performance (initial hypothesis later refuted).
  H2: Financial risk mediates the ESGP–firm performance relationship.
  H3a: In the growth phase, financial risk does not mediate the ESGP–performance relationship.
  H3b: At the maturity stage, financial risk mediates the ESGP–performance relationship.
  H3c: In the decline phase, financial risk does not mediate the ESGP–performance relationship.

Empirical findings (confirmation or not of the hypotheses):

H1: Refuted. ESGP is positively related to performance (coeff. ≈ 0.150, $p < 0.01$).
H2: Confirmed. Financial risk (Z-score) positively mediates the ESGP–performance relationship.
H3a: Refuted. Risk mediates the ESGP–performance relationship even in growth phases.
H3b: Confirmed. Significant mediation in maturity.
H3c: Confirmed. No significant mediation in decline.

Moderating effects:

Quality Disclosure: Strengthens the positive effect of ESG on performance.
Financial Slack: Reduces the Effectiveness of ESG in the Maturity Phase.

Environmental uncertainty weakens the effectiveness of ESG, especially in mature and declining phases.

Research Method: (i) Fixed effects regression (industry-year); (ii) Mediation model (Baron & Kenny, 1986); (iii) Instrumentation (2SLS) and regression with lagged variables; (iv) Moderation analysis (interactions); (v) Robustness test with alternative ESG and alternative performance (ROA, ROE).

Period: 2010–2020.

Sample & Data:

1486 Chinese A-share enterprises.
11,742 firm-year observations.
Sources: Wind (Huazheng ESG rating), CSMAR.

Variables & Controls:

Dependent:

- Tobin's Q (main).
- ROA, ROE (robustness test).

Explanatory:

- ESG score (Huazheng, scale 1–9).
- ESG_H (Rankins, for robustness).
- ESG^2 for non-linearity tests (positive U-shaped relationship).

Mediator:

- Z-score (Altman): financial risk.

Moderator:

- Disclosure quality (A–D rating).
- Financial slack (equity/liquidity).
- Environmental uncertainty (sales standard deviation).

Controls:

- Size (total active logs).
- SOE (dummy state-owned enterprises).
- Age (business age).
- Leverage.
- Growth (change in revenues).
- Institutional investors' shareholding.
- Ownership concentration (top 10 shareholders).

06.

Title: Does it pay to deliver superior ESG performance? Evidence from US S&P 500 companies.

Authors: Bejtush Ademi, Nora Johanne Klungseth.

Published: 2022 (Journal of Global Responsibility, 13(4), pp. 421–449).

URL: https://doi.org/10.1108/JGR-01-2022-0006.

Summary: The study investigates the impact of ESG performance on the financial return (ROCE) and market value (Tobin's Q) of 150 US companies listed in the S&P 500 over the period 2017–2020, including the Covid-19 crisis. The analysis shows that companies with superior ESG performance achieve better financial results and are more favourably valued by the market compared to their peers. The evidence supports the usefulness of ESG metrics for investors and companies even during unexpected crises.

Theoretical framework: The article adopts a multilevel theoretical approach, mainly based on four key theories: (i) Stakeholder Theory (Freeman, 2010). It is the main theoretical framework adopted to justify the thesis according to which positive ESG performance builds trust and lasting relationships with stakeholders, generating value even in times of crisis (e.g., Covid-19). According to this perspective, companies that responsibly manage stakeholder relationships are rewarded by the market; (ii) Resource-Based View (RBV): ESG is interpreted as a source of sustainable competitive advantage. ESG investments contribute to the development of intangible corporate resources and capabilities (such as reputation, operational efficiency and risk management), with positive impacts on financial performance and market value; (iii) Value Creation Theory: ESG practices are seen as strategic levers for creating value in the long term, through product differentiation, reputation, customer loyalty and reduction of the cost of capital; (iv) Cost of Capital Reduction Theory: ESG reduces the risk perceived by the company (regulatory, reputational, operational), determining a lowering of the cost of capital and therefore an increase in enterprise value. It is a central element to explain the positive relationship between ESG and Tobin's Q. The authors implicitly reject the Shareholder Theory (Friedman, 2007), criticising the idea that the only objective of the company is the maximisation of profit. On the contrary, they argue that ESG investments, although sometimes expensive in the short term, are profitable in the medium-long term.

Research Questions & Hypotheses:

- Research Questions:

  Does ESG performance improve corporate financial results?
  Does ESG performance increase the market valuation of companies?

- *Hypotheses*:

  H1: ESG performance has a significant positive impact on financial performance (ROCE).
  H2: ESG performance has a significant positive impact on market value (Tobin's Q).

  Empirical findings (confirmation or not of the hypotheses):

H1: Confirmed. The WLS model shows a positive and significant effect of ESG on ROCE ($\beta = 0.015$, $p < 0.05$).
H2: Confirmed. ESG has a positive and significant impact on Tobin's Q ($\beta = 0.203$, $p < 0.01$), even during the Covid-19 crisis.

Research Method: (i) Regression with panel data: fixed effects and Weighted Least Squares (WLS) models; (ii) Hausman test for choosing between fixed and random effects; (iii) Heteroskedasticity test and correction with WLS; (iv) Variables collected from MSCI, corporate balance sheets, and Yahoo Finance.
Period: 2017–2020 (includes the Covid-19 crisis).
Sample & Data:

150 companies in the S&P 500.
600 firm-year observations.
ESG ratings from MSCI (scale 1–7: CCC = 1, AAA = 7).
Sectors represented: 50.
Financial data from company reports and Macrotrends.

Variables & Controls:

Dependent:

- ROCE (Return on Capital Employed).
- Tobin's Q (market value).

Explanatory:

- ESG Rating (MSCI score, scale 1–7).

Controls:

- Debt-to-equity ratio (DOT).
- Company size (total active logs).

Others:

- Years included: 2017, 2018, 2019, 2020.
- Covid-19 crisis considered in the model.
- Graphical analysis of ESG distributions over time.

07.

Title: The effect of ESG and CSR attitude on financial performance in Europe: A quantitative re-examination.

Author: Paolo Candio.

Published: 2024 (Journal of Environmental Management, 354, 120390).

URL: https://doi.org/10.1016/j.jenvman.2024.120390.

Summary: The study examines the association between ESG performance and financial performance in a sample of listed European companies, and analyses whether and how CSR attitude moderates this relationship. Six financial performance indicators and three regression frameworks are used, with attention to sector effects through mixed-effects models. The results highlight a high heterogeneity of ESG effects, a weak significance of the moderation by CSR attitude, and a relevant impact of the methodological choice on the empirical outcomes.

Theoretical Framework: The article is built on two main theoretical references: (i) Value-Relevance Theory (Barth et al., 2001): This theory interprets accounting information (and, by extension, ESG metrics) as relevant for the determination of the market value of companies. The underlying hypothesis is that companies that provide quality ESG disclosures, integrating sustainable practices in daily management, are positively perceived by investors and can obtain competitive advantages, such as better reputation and better financial performance; (ii) Neo-Institutional Theory (DiMaggio & Powell, 1991; Scott, 2001): According to this perspective, companies tend to conform to institutional, normative and mimetic pressures in their behaviour, including that related to sustainability. The presence of CSR committees, external verification of reports, and stakeholder engagement are seen as institutionalised responses to such pressures. The article applies this framework to analyse variations in the effectiveness of ESG depending on the CSR attitude and the sector context. Furthermore, Stakeholder Theory is referred to as a supporting framework, insofar as it is assumed that companies improve their performance through more responsible management towards all stakeholders, but this theory does not constitute the main analytical pillar of the study.

Research Questions & Hypotheses:

- Research Questions:

  What is the relationship between ESG scores and the financial performance of European companies?
  Does the firm's attitude toward CSR moderate this relationship?

- Hypotheses:

  H1: ESG performance is positively associated with financial performance.
  H2: Attitude toward CSR moderates the ESG–financial performance relationship.

Empirical findings (confirmation or not of the hypotheses):

H1: Partially confirmed: (i) ESG score negatively associated with ROA ($\beta = -0.0793$, $p < 0.01$); (ii) Non-significant association with other aggregate indicators; (iii)

The effects of the individual ESG pillars are heterogeneous and vary by indicator (e.g. Environmental score positive for share price but negative for ROA).

H2: Partially confirmed: (i) The presence of a CSR committee negatively moderates the ESG–ROA relationship ($\beta = -0.0637$, $p < 0.05$); (ii) External auditing of the CSR report negatively moderates the ESG–share price relationship ($\beta = -2.7967$, $p < 0.01$); (iii) The other CSR attitude variables do not show significant effects.

Research Method: (i) Multilevel regression (GLMM) with mixed effects (sector clustering); (ii) OLS regression and panel regression for robustness comparison; (iii) ESG × CSR attitude interactions; (iv) Hausman test for model choice; (v) Sensitivity analysis on model specification.

Period: 2012–2022.

Sample & Data:

600 companies are listed in the STOXX Europe 600.
6600 firm-year observations.
Sources: Refinitiv Eikon for ESG, financial indicators, and CSR variables.
Software: STATA 16.

Variables & Controls:

Dependent (financial indicators):

- ROA, ROE, EBIT margin, EPS, P/E ratio, Share Price.

Explanatory:

- Total ESG score (0–100).
- Environmental, Social, Governance scores.

Moderating (CSR attitude):

- CSR Strategy (score 0–100).
- Stakeholder Engagement (dummy).
- CSR External Audit (dummy).
- CSR Committee (dummy).

Controls:

- Size: log market value, total assets, revenues, capitalisation.
- Leverage, Beta (volatility), Book Value per Share, Market-to-Book Value.

08.

Title: ESG Performance and Corporate Resilience: An Empirical Analysis Based on the Capital Allocation Efficiency Perspective.

Authors: Kun Wang, Shidong Yu, Mingshuai Mei, Xiangyang Yang, Guanyu Peng, Bing Lv.

Published: 2023 (Sustainability, 15(23), Art. 16145).

URL: https://doi.org/10.3390/su152316145.

Summary: The study examines the impact of ESG performance on corporate resilience, measured through capital allocation efficiency. The hypothesis that firms with high ESG performance are more resilient because they are able to maintain or improve allocative efficiency even in the presence of exogenous shocks is tested. The model also explores the role of ownership concentration and institutional quality (at the provincial level) as moderating variables. The results show that ESG significantly improves allocative resilience, with more marked effects in areas with high institutional quality and in firms with dispersed ownership.

Theoretical framework: The article is based on an integrated theoretical framework, divided into three fundamental references: (i) Stakeholder Theory (Freeman, 2010): It considers ESG as a lever to improve stakeholder trust, reduce information asymmetry, and lower transaction costs. Companies that adopt ESG practices send positive signals to the market, obtaining greater support from investors, customers and other stakeholders, which favours corporate resilience; (ii) Resource Dependence Theory (Pfeffer & Salancik, 1978): ESG is interpreted as a tool to improve access to strategic resources, thanks to the strengthening of relationships with key stakeholders. Resilient companies are those that are able to attract and maintain critical resources (financial, human, relational) even in shock situations; (iii) Signalling Theory: Good ESG performance and their transparent communication are seen as signals of quality and reliability. The adoption of ESG standards, especially in emerging contexts, helps differentiate the company, reducing the uncertainty perceived by investors.

Research Questions & Hypotheses:

- Research Questions:

  Does ESG performance improve corporate resilience through greater capital allocation efficiency?

  Do institutional quality and ownership structure influence this relationship?

- Hypotheses:

  H1: ESG performance improves capital allocation efficiency, contributing to corporate resilience.

  H2: The ESG-resilience relationship is strengthened by low ownership concentration.

  H3: Institutional quality positively moderates the ESG-Resilience relationship.

Empirical findings (confirmation or not of the hypotheses):

H1: Confirmed. ESG is positively and significantly associated with allocative efficiency (coeff. ≈ 0.0156, $p < 0.01$).

H2: Confirmed. The positive effect of ESG is stronger in firms with low ownership concentration.

H3: Confirmed. Institutional quality at the local (provincial) level strengthens the ESG-resilience relationship.

Research Method: (i) Panel regressions with fixed effects (industry and year); (ii) Moderating interactions; (iii) Robustness tested with alternative ESG variables and GMM method; (iv) Instrumental variables: local shocks and sectoral ESG average.

Period: 2011–2021.
Sample & Data:

2613 Chinese non-financial A-share enterprises.
Sources: CSMAR (financial data), WIND (ESG), provincial institutional databases.
Total observations: approximately 21,000 firm-years.

Variables & Controls:

Dependent (resilience proxy):

- Allocative efficiency of capital, measured with the Richardson (2006) model based on expected optimal investments.

Explanatory:

- ESG Score (WIND ESG rating).

Moderator:

- Concentrated ownership (dummy: top 10 shareholders > 50%).
- Regional institutional quality index.

Controls:

- Firm size (log total assets).
- Leverage.
- Profitability (ROA).
- Sales growth.
- Asset tangibility.
- Board independence.
- Industry and year fixed effects.

09.
Title: Investigating the effect of ESG disclosure on firm performance: The case of Saudi Arabian listed firms.
Authors: Egi Arvian Firmansyah, Umar Habibu Umar, Rabiu Saminu Jibril.
Published: 2023 (Cogent Economics & Finance, 11(2), Art. 2287923).
URL: https://doi.org/10.1080/23322039.2023.2287923.
Summary: The study analyses the impact of ESG disclosure on the financial performance (measured by ROA and ROE) of listed companies in Saudi Arabia, also considering the effect of company size as a moderating variable. The results show that ESG disclosure significantly improves profitability and that larger companies benefit more from ESG practices. The analysis is part of the Vision 2030 reforms, which push towards greater sustainability in the Saudi economic system.
Theoretical framework: The article is based on three main theoretical references: (i) Stakeholder Theory (Freeman, 2010): It is the central theoretical pillar. It argues that the adoption of ESG practices and their transparent communication improves the relationship with stakeholders (customers, suppliers, investors, communities),

promoting long-term sustainable performance; (ii) Slack Resources Theory (Waddock & Graves, 1997): Companies with favourable financial performance have available resources (slack) to invest in ESG initiatives. The adoption of ESG is therefore a consequence of abundance of resources, and not necessarily a direct cause of financial performance; (iii) Agency Theory (Jensen & Meckling, 1976)—recalled in contrast: According to the Agency, ESG activities can lead to an inefficient use of resources if motivated by managerial interests, resulting in a loss of value for shareholders. The authors critically discuss it, underlining how the evidence on the Saudi context challenges this view.

Research Questions & Hypotheses:

- Research Questions:

  Does ESG disclosure have a positive impact on the financial performance of Saudi firms?

  Does company size moderate this relationship?

- Hypotheses:

  H1: ESG disclosure is positively associated with financial performance (ROA and ROE).

  H2: Firm size positively moderates the ESG–financial performance relationship.

Empirical findings (confirmation or not of the hypotheses):

H1: Confirmed. ESG disclosure has a positive and statistically significant effect on both ROA ($\beta = 0.183$, $p < 0.01$) and ROE ($\beta = 0.201$, $p < 0.05$).

H2: Confirmed. The ESG disclosure × firm size interaction is positive and significant, indicating that larger firms obtain greater benefits in terms of financial performance from ESG disclosure.

Research Method: (i) Multiple linear regression; (ii) Moderation analysis through interactions (ESG × firm size); (iii) Variable normalisation with natural logarithm; (iv) Check for multicollinearity and diagnostic tests (VIF).

Period: 2017–2021.

Sample & Data:

100 Saudi companies listed on the Saudi Exchange (Tadāwul).
Excluded sectors: finance and insurance.
Data collected from annual reports and the Refinitiv database.
Total observations: 500 firm-years.

Variables & Controls:

Dependent:

- ROA.
- ROE.

Explanatory:

- ESG Disclosure Score (scale 0–1, taken from Refinitiv)

Moderator:

- Firm Size (log total assets).

Controls:

- Leverage (debt/asset).
- Asset Turnover.
- Firm Age.
- Board Size.
- Industry dummies.

10.

Title: Environmental, Social and Governance (ESG) Scores and Financial Performance of Multilatinas: Moderating Effects of Geographic International Diversification and Financial Slack.

Authors: Eduardo Duque-Grisales, Javier Aguilera-Caracuel.

Published: 2021 (Journal of Business Ethics, 168(2), pp. 315–334).

URL: https://doi.org/10.1007/s10551-019-04177-w.

Summary: The study analyses the effect of ESG performance on the financial performance of a sample of 98 Latin American multinational companies (multilatinas) in an emerging economies context and evaluates the moderating role of international geographic diversification (GID) and financial slack. The results show that ESG, alone, does not directly improve performance; however, in the presence of high GID and financial slack, the ESG-performance link becomes positive and significant. The article suggests that ESG investments are effective only in combination with adequate resources and strategies.

Theoretical framework: The article is based on a dual theoretical framework, articulated in two main theories: (i) Resource-Based View (RBV): The RBV approach is used to explain how firms (multilatinas) that adopt ESG practices develop distinctive organisational resources and capabilities, such as reputation, management efficiency, and stakeholder relations—that can translate into competitive advantage in the long run. However, in emerging economies such as Latin American ones, scarce resources and inefficiency in ESG processes can reverse this link, making such practices costly and ineffective; (ii) Institutional Theory (Kostova & Zaheer, 1999): This theory is applied to analyse the influence of the institutional context on ESG practices and performance. Firms operating in weak institutional environments, such as many Latin American countries, face greater legitimacy pressures when they internationalise. Geographic International Diversification (GID) thus becomes a mechanism to improve reputation and adherence to ESG standards, strengthening financial performance. The authors also incorporate the concept of Financial Slack, understood as an available financial resource that can enable more strategic and visible ESG investments, acting as a moderator in the ESG-performance relationship.

Research Questions & Hypotheses:

- Research Questions:

  Does ESG Improve the Financial Performance of Internationalised Latin American Companies?

  Do the degree of international geographic diversification (GID) and financial slack moderate this relationship?

- Hypotheses:

  H1: ESG is positively correlated with financial performance (ROA).

  H2: Geographic diversification (GID) positively moderates the ESG-performance relationship.

  H3: Financial slack positively moderates the ESG-performance relationship.

Empirical findings (confirmation or not of the hypotheses):

H1: Not confirmed. The relationship between ESG and ROA is not significant in general.

H2: Confirmed. The GID moderator makes the ESG–ROA relationship significant and positive.

H3: Confirmed. Financial slack also acts as a positive moderator.

Research Method: (i) OLS regression with interactions (moderators); (ii) Control for sector and time effects; (iii) Robustness tested with alternative models (ROE, total assets).

Period: 2011–2016.

Sample & Data:

98 multilatinas (multinational companies based in Latin America).
486 firm-year observations.
Main countries: Brazil, Mexico, Argentina, Chile, Colombia.
ESG from Thomson Reuters.
Financial data from Osiris and Bloomberg.

Variables & Controls:

Dependent:

- ROA (Return on Assets).

Explanatory:

- ESG Score (Thomson Reuters—Refinitiv).

Moderator:

- Geographic International Diversification (n. foreign countries operating).
- Financial Slack (current assets/current liabilities).

Controls:

- Firm size.
- Age.
- Leverage.
- Capital intensity.
- Industry and year dummies.

11.

Title: ESG Performance and Disclosure: A Cross-Country Analysis.

Authors: Florencio Lopez-de-Silanes, Joseph A. McCahery, Paul C. Pudschedl.

Published: 2020 (Singapore Journal of Legal Studies, March 2020, pp. 217–241).

URL: https://www.jstor.org/stable/10.2307/27032607.

Summary: The study analyses the determinants and effects of ESG performance and ESG disclosure on an international scale, comparing 30 countries. The authors explore the role of ESG regulation, corporate governance characteristics, and market dynamics in fostering transparency. The results indicate that the quality of ESG disclosure is positively associated with the quality of ESG performance and the local regulatory framework. However, the relationship between ESG and financial performance varies significantly based on the regulatory context and the presence of mandatory disclosure.

Theoretical framework: The paper adopts an integrated theoretical approach that combines elements from financial economics, risk theory and information theory: (i) Signalling Theory: Firms with high ESG quality and high transparency in reporting signal lower risk and higher management quality, attracting investors willing to pay a lower return in exchange for less uncertainty. The paper assumes that high ESG disclosure can reduce the cost of capital; (ii) Investors' non-financial utility theory: Some investors incorporate a non-purely financial component (e.g., ethical, reputational) in their selection criteria, accepting lower returns in exchange for investing in firms with good ESG performance; (iii) Adverse selection theory: In markets without mandatory ESG disclosure, only firms with good ESG indicators have an incentive to disclose information. This creates a positive relationship between quality and quantity of disclosure in less regulated contexts (e.g., USA); (iv) Systemic risk and resilience theory: ESG is seen as a risk reduction mechanism (e.g., reputational, regulatory, operational), with positive effects on return volatility. However, this does not always translate into better risk-adjusted performance.

Research Questions & Hypotheses:

- Research Questions:

  What factors explain the quality of ESG disclosure across countries?

  How does the relationship between ESG and financial performance vary with transparency and regulation?

  Is ESG disclosure associated with risk-adjusted performance?

- Hypotheses:

H1: The quality of ESG disclosure is higher in countries with less regulatory obligation, where adverse selection is at work.
H2: ESG and disclosure are positively associated with risk-adjusted performance in less-regulated markets.
H3: The ESG-performance relationship is mediated by the legal, cultural, and regulatory context.

Empirical findings (confirmation or not of the hypotheses):

H1: Confirmed. In countries with voluntary disclosure (e.g., USA), companies that disclose voluntarily have higher ESG.
H2: Partially confirmed. ESG is associated with lower volatility and higher resilience, but not necessarily with higher financial returns.
H3: Confirmed. The strength of law, the level of enforcement, and national culture (e.g., individualism, distance from power) significantly influence ESG disclosure and its effectiveness.

Research Method: (i) Cross-country analysis with multiple regressions; (ii) Panel regression on firm-year data (≈10,000 firms), (iii) Aggregated indicators from Bloomberg, Refinitiv, Sustainalytics; (iv) Legal and cultural variables from World Bank, Hofstede, WGI.
Period: 2014–2018.
Sample & Data:

9941 listed companies in 30 countries.
Diversified sectors.
ESG scores and disclosure from Bloomberg ESG, Refinitiv.
Institutional and cultural variables from the World Bank, WGI, and Hofstede.

Variables & Controls:

Dependent:

- Risk-adjusted performance (Sharpe ratio, volatilities).
- ESG disclosure quality.
- ESG performance score.

Explanatory:

- Mandatory disclosure (dummy).
- ESG score.
- Institutional variables (e.g., Rule of Law, Voice & Accountability).
- Cultural factors (Hofstede: individualism, hierarchical distance).

Controls:

- Size.
- Industry.
- Leverage.

- Board independence.
- CEO duality.
- Profitability.
- Country fixed effects.

12.

Title: Does ESG Disclosure Influence Firm Performance?

Authors: Silvia Carnini Pulino, Mirella Ciaburri, Barbara Sveva Magnanelli, Luigi Nasta.

Published: 2022 (Sustainability, 14(13), Art. 7595).

URL: https://doi.org/10.3390/su14137595.

Summary: The paper analyses whether and to what extent ESG disclosure affects corporate performance, with particular attention to the Italian context. Using a sample of 94 listed Italian companies, the paper investigates the link between the degree of ESG disclosure and three performance indicators (ROA, ROE, and Tobin's Q). The results confirm the existence of a positive and statistically significant relationship between ESG disclosure and performance, especially in terms of market value (Tobin's Q). The paper fits into the literature on transparency, underlining the importance of ESG reporting for value creation.

Theoretical framework: The article adopts a theoretical framework based on two key theories: (i) Agency Theory (Jensen & Meckling, 1976): ESG disclosure reduces the information asymmetry between management (agent) and shareholders (principal), helping to contain agency costs. Greater transparency is interpreted as a form of alignment of interests between managers and investors; (ii) Signalling Theory (Spence, 1973): ESG disclosure acts as a positive signal that companies send to stakeholders to communicate commitment and reliability. High disclosure is seen as evidence of good governance, attention to social and environmental impacts, and long-term orientation. These two theories are used to justify the positive effect of ESG disclosure on corporate performance: reduction of uncertainty, increase of stakeholder trust, and incentive to virtuous behaviour by managers.

Research Questions & Hypotheses:

- Research Questions:

  Does ESG disclosure positively influence corporate performance?
  Is the relationship also significant in the Italian context, characterised by evolving ESG regulations?

- Hypotheses:

  H1: There is a positive relationship between ESG disclosure and corporate performance (ROA, ROE, Tobin's Q).

Empirical findings (confirmation or not of the hypotheses):

H1: Partially confirmed.
Tobin's Q: positive and significant relationship ($\beta = 0.313$, $p < 0.05$).
ROA: positive but not significant relationship.

ROE: non-significant relationship.
Conclusion: ESG disclosure positively influences market value but does not significantly impact accounting profitability.

Research Method: (i) Multiple linear regression; (ii) Cross-section analysis; (iii) Sample selected from Italian companies listed on the FTSE MIB; (iv) ESG disclosure score constructed according to 6 areas (ex-Legislative Decree 254/2016).
Period: 2011–2020.
Sample & Data:

263 listed Italian companies.
ESG data is manually collected from financial statements and sustainability reports.
Construction of a composite indicator of ESG disclosure (score 0–1) based on six macro-areas: Business model, Sustainability policies, ESG risks, ESG objectives and results, Key Indicators (KPI), Accounting principles used in non-financial reporting.

Variables & Controls:

Dependent:

- ROA.
- ROE.
- Tobin's Q.

Explanatory:

- ESG Disclosure Score (built on 6 areas, scale 0–1).

Controls:

- Size (log total assets).
- Leverage (debt/asset).
- Sector dummies.

13.
Title: Does ESG Performance Have an Impact on Financial Performance? Evidence from Turkey.
Authors: Melih Sefa Yavuz, Hasan Sadık Tatlı, Gözde Bozkurt, Gökten Öngel
Published: 2025 (Journal of Entrepreneurship, Management and Innovation, 21(1), pp. 24–42).
URL: https://doi.org/10.7341/20252112.
Summary: The study analyses the impact of three ESG performance dimensions on the financial performance of 21 Turkish firms listed in the BIST 100, over the period 2011–2020. The authors use two performance measures: ROE and Tobin's Q (market value proxy). The results show that only the Governance dimension has a positive and significant effect on both variables, while the Environmental and Social components do not significantly influence the firm's performance.
Theoretical framework: The article is based on a complex theoretical framework, which includes three main interpretative hypotheses drawn from the literature on

CSR and financial performance: (i) Stakeholder Theory (Freeman, 1994): It is the main theoretical basis. According to this theory, companies that respond to the needs of all stakeholders (and not only shareholders) create sustainable value in the long term. The authors believe that good ESG performance can consolidate trust, reputation and relational ties with stakeholders, with benefits on profitability and market value; (ii) Social Impact Hypothesis (Waddock & Graves, 1997): This hypothesis suggests that high ESG performance produces positive effects on financial performance through greater stakeholder satisfaction and better risk management; (iii) Negative Synergy & Balancing Hypothesis (Mcwilliams & Siegel, 1997): Alternative hypotheses that see ESG investments as potentially detracting from financial performance are also discussed. According to these views, ESG initiatives can absorb resources, reduce profits, and do not generate immediate competitive advantages, especially in emerging or poorly regulated contexts. The article uses these theories to formulate and test hypotheses regarding the impact of environmental, social, and governance dimensions on the financial performance of Turkish companies listed in the Borsa Istanbul-100 index. The authors do not take a priori positions but empirically test which of the theoretical hypotheses is more supported by the data. The results, which show a positive effect only for Governance, partially confirm the Stakeholder Theory and the Social Impact Hypothesis but also offer critical insights for the other ESG pillars.

Research Questions & Hypotheses:

- Research Questions:

  Do the three ESG pillars influence the profitability (ROE) and market value (Tobin's Q) of Turkish companies?

  Which ESG component is most relevant to financial performance?

- Hypotheses:

  H1: There is a significant relationship between environmental performance (E) and financial performance (ROE, Tobin's Q).

  H2: There is a significant relationship between social performance (S) and financial performance.

  H3: There is a significant relationship between governance performance (G) and financial performance.

Empirical findings (confirmation or not of the hypotheses):

H1: Not confirmed. No significant effect of Environment size on ROE or Tobin's Q.

H2: Not confirmed. No significant relationship between Social and the two performance measures.

H3: Confirmed. Governance has a positive and significant impact on ROE and Tobin's Q.

Research Method: (i) Panel regression with fixed effects; (ii) Hausman test for model choice; (iii) Separate analyses on ROE and Tobin's Q as dependent variables.

Period: 2011–2020.

Sample & Data:

21 Turkish companies are listed in the BIST 100.
376 firm-year observations.
Sources: Refinitiv Eikon for ESG and accounting data.

Variables & Controls:

Dependent:

- ROE.
- Tobin's Q (market value/book value).

Explanatory:

- Environmental Score (E).
- Social Score (S).
- Governance Score (G).

Controls:

- Firm size (log total assets).
- Leverage (debt/asset).
- Firm age (age).
- Industry dummies.

14.
Title: Variations in Financial Performance of Firms with ESG Integration in Business: The Mediating Role of Corporate Efficiency Using DEA.

Authors: Abhisek Mahanta, Naresh Chandra Sahu, Pradeep Kumar Behera, Pushp Kumar.

Published: 2024 (Green Finance, 6(3), pp. 518–562).

URL: https://doi.org/10.3934/GF.2024020.

Summary: The study analyses how the integration of ESG practices impacts Corporate Efficiency (CE), measured through Data Envelopment Analysis (DEA), and how such efficiency mediates the effect of sustainability on the financial performance (CFP) of Indian firms. The analysis considers both the aggregate ESG and its three pillars (E, S, G), highlighting that firms with higher ESG performance tend to be more efficient and, consequently, better performing from an economic-financial perspective.

Theoretical framework: The article is based on an integrated theoretical framework, which combines three complementary interpretative approaches: (i) Stakeholder Theory (Freeman, 2010): Companies are not only accountable to shareholders, but to all stakeholders (customers, employees, communities, institutions) → The adoption of ESG practices is interpreted as a strategic response to social and environmental expectations, useful to generate trust, consensus and long-term sustainability; (ii) Legitimacy Theory (Dowling & Pfeffer, 1975; Suchman, 1995): According to this theory, organisations seek to obtain and maintain social legitimacy through behaviours that are perceived as appropriate by the regulatory and cultural context → ESG integration is seen as a tool to strengthen legitimacy, improve public image and prevent conflicts with stakeholders; (iii) Resource-Based View (RBV) (Barney, 1991): ESG capabilities are considered strategic intangible assets, difficult to imitate, that can improve operational efficiency and provide sustainable competitive advantages → The study adopts Corporate Efficiency (CE) as a mediating variable between ESG and financial performance (CFP): ESG → Efficiency (CE) → Financial Performance (ROA, ROE). The study combines the ethical-legitimating view (Stakeholder & Legitimacy Theory) with the strategic-efficiency view (RBV), arguing that the adoption of ESG practices is an effective lever to improve performance by strengthening operational efficiency.

Research Questions & Hypotheses:

- Research Questions:

  Does ESG Integration Improve Corporate Financial Performance?
  Is the ESG-performance link mediated by corporate efficiency?
  Do efficient firms outperform inefficient ones?

- Hypotheses:

  H1: Aggregate ESG performance has a positive influence on corporate efficiency.
  H2: The environmental pillar (E) has a positive influence on corporate efficiency.
  H3: The social pillar (S) has a positive influence on corporate efficiency.
  H4: The governance pillar (G) has a positive influence on corporate efficiency.
  H5: Environmental activities have a positive influence on business efficiency.

Empirical findings (confirmation or not of the hypotheses):

H1–H5: All confirmed. The overall ESG scores and the three components (E, S, G) show positive and significant effects on corporate efficiency. The most efficient companies, identified through DEA, show better financial results in terms of ROA, ROE, and market capitalisation.

Research Method: (i) DEA (Data Envelopment Analysis): to calculate the technical efficiency (input-output); (ii) Panel regression (OLS with fixed effects): ESG → CE → CFP; (iii) Mediation analysis with stepwise approach; (iv) T-test to compare groups (efficient vs. inefficient).

Period: 2022.

Sample & Data:

909 listed non-financial companies in India.
Data from Refinitiv, ACE Equity, and the RBI Database.
909 firm-year observations.

Variables & Controls:

Dependent:

- ROA.
- ROE.
- Market Capitalisation (alternative proxy of CFP).

Explanatory:

- Aggregate ESG Score.
- ESG pillars (E, S, G).

Mediator:

- Corporate Efficiency (calculated with DEA, input: assets, capital, expenses; output: revenue, profit, share price).

Controls:

- Size (log total assets).
- Firm age.
- Leverage.
- Industry dummies.
- Fixed year effects.

15.

Title: The Impact of ESG Practices on Firm Performance: Evidence from Emerging Countries.

Authors: Muhammad Naeem, Hamid Ullah, Shahid Jan.

Published: 2021 (Indian Journal of Economics and Business, 20(1), pp. 731–750).

URL: http://www.ashwinanokha.com/IJEB.php.

Summary: This study explores the effect of ESG (Environmental, Social, and Governance) practices on the financial performance of companies belonging to emerging markets (China, Brazil, South Africa, and Russia). Using a sample of 1042 listed companies for the period 2010–2019, the study examines whether higher ESG performance can translate into better financial performance, as measured by indicators such as ROA (Return on Assets) and Tobin's Q (market performance). The results indicate that ESG practices overall significantly improve financial performance, highlighting differentiated effects based on the ESG pillars analysed individually.

Theoretical framework: The article mainly uses Stakeholder Theory as its main theoretical framework: (i) Stakeholder Theory (Freeman, 2010). According to this theoretical approach, companies must satisfy the demands of all stakeholders (not only shareholders). The authors adopt this theory to explain that ESG practices represent concrete efforts by management to respond to the expectations of internal and external stakeholders. Stakeholder theory indeed predicts that the adoption of ESG practices improves stakeholder trust, reduces corporate risk, and, consequently, increases market value and corporate profitability. The authors also discuss several empirical results from previous literature, highlighting several contrasting positions regarding the effects of ESG on financial performance, distinguishing between positive, negative, or null effects. However, the central framework adopted remains focused on Stakeholder Theory, used to theoretically motivate the expectation of a positive impact of ESG practices on corporate performance, measured through Tobin's Q and ROA.

Research Questions & Hypotheses:

Research Questions:

What is the overall impact of ESG practices on the financial performance of companies in emerging markets?
Which ESG (environmental, social, governance) components most influence corporate performance?

Hypotheses:

H1: Overall, ESG practices have a significant positive impact on the financial performance of companies in emerging markets.
H2: The environmental pillar (Environmental) has a significant positive impact on financial performance.
H3: The social pillar (Social) has a significant positive impact on financial performance.
H4: The governance pillar (Governance) has a significant positive impact on financial performance.

Empirical findings (confirmation or not of the hypotheses):

H1: Confirmed. ESG integration has a significant positive impact on both ROA and Tobin's Q.
H2: Confirmed. The environmental component has a significant positive impact on Tobin's Q and ROA.
H3: Confirmed. The social dimension positively and significantly influences Tobin's Q and ROA.
H4: Partially confirmed. The governance dimension has a significant positive impact on Tobin's Q, while it is less strong (still significant) on ROA.

Research Method: (i) Panel data analysis (fixed effects model), (ii) Hausman test for choosing the appropriate model (fixed vs random effects); (iii) Controls for multicollinearity, heteroscedasticity, and autocorrelation using appropriate diagnostic tests.

Period: 2010–2019.

Sample & Data:

1042 listed companies in four emerging countries: China, Brazil, South Africa, and Russia.
6280 firm-year observations.
ESG and financial performance data source: Thomson Reuters (Refinitiv).

Variables & Controls:

Dependent:

- ROA (Return on Assets).
- Tobin's Q (market performance indicator).

Explanatory:

- ESG combined score (aggregate).
- Environmental Score (E).
- Social Score (S).
- Governance Score (G).

Controls:

- Company size (log Total Assets).
- Leverage (Debt-to-Assets).
- Sales Growth.
- Firm Age.
- Dummies by country and sector.

6.2 ESG and Credit Rating: Mechanisms and Effects

The relationship between ESG factors and corporate debt ratings represents an area of growing relevance in the academic literature and the risk assessment practices adopted by banks and rating agencies. With the intensification of attention towards sustainability and responsible corporate management, credit analysts are progressively integrating ESG pillars into their assessments of corporate credit standing. The underlying hypothesis is that a strong ESG performance can reduce the overall perceived risk of the company, positively influencing access to credit and financing conditions. This review examines the empirical literature published between 2020 and 2025 in scientific journals indexed by Scopus and Web of Science, with the aim of analysing the main findings, research questions, methodologies employed, and variables considered in the studies investigating the impact of ESG indicators on corporate credit ratings.

01.
Title: The impact of ESG factors on credit ratings: An empirical study of European banks.

Authors: Sheng-Li Chen, Dan-ni Wang, Hao-Wei Chen.

Published: 2025 (International Review of Economics & Finance, 99, 104056).

URL: https://www.sciencedirect.com/science/article/pii/S1059056025002199.

Summary: The study analyses the impact of ESG factors on the long-term debt ratings assigned by Moody's, S&P, and Fitch to 106 European listed banks. The results indicate that environmental factors exert a significant positive influence on ratings by all three agencies. However, the effect of overall ESG, social, governance, and ESG controversies scores varies across rating agencies, confirming the methodological inconsistency and difficulty of standardisation in ESG assessment. The work highlights the importance of greater transparency and consistency in ESG risk measurement for a more effective integration into credit assessment processes.

Theoretical framework: The paper is mainly based on stakeholder theory (Freeman, 1994), according to which the creation of long-term sustainable value passes through the effective management of relationships with all stakeholders. Agency theory is also considered (Jensen & Meckling, 1976), according to which ESG investments can represent a cost not aligned with shareholders' interests. The hypotheses formulated reflect the stakeholder-oriented theoretical approach.

Research Questions & Hypotheses:

Research Questions:

To what extent do different dimensions of ESG performance influence the long-term debt ratings assigned by Moody's, S&P, and Fitch to European banks?
Are there systematic differences between rating agencies in integrating ESG factors into their rating models?

Hypotheses:

H1: The ESG Combined Score positively influences banks' ratings.
H2: The environmental dimension score (Environmental Pillar Score) positively influences banks' ratings.
H3: The Social Pillar Score has a positive impact on bank ratings.
H4: The governance dimension score (Corporate Governance Pillar Score) positively affects banks' ratings.
H5: ESG Controversies Score positively influences banks' ratings.

Empirical findings (confirmation or not of the hypotheses):

H1: Confirmed only for Fitch. No significant effect on Moody's and S&P.
H2: Confirmed for all three agencies (significant and positive).
H3: Rejected. The Social Pillar Score has a significant negative impact only on S&P, not on Moody's and Fitch.
H4: Partially confirmed. Significant and positive only for Fitch.
H5: Partially confirmed. Significant and positive only for S&P.

Research Method: Mixed-effects ordered logistic regression model, with lagged independent and control variables to mitigate endogeneity issues.

Period: 2019–2023.

Sample & Data:

106 European-listed banks from 26 countries.
ESG, financial, and rating data collected from the Refinitiv database (LSEG).
Ratings by Moody's, S&P, and Fitch.

ESG scores: ESG Combined, Environmental, Social, Governance, Controversies (Refinitiv methodology).

Variables & Controls:

Dependent:

- Long-term debt rating ranking (ordinal variable from 0 = lowest rating to 10 = highest rating).

Explanatory:

- ESG Combined Score.
- Environmental Pillar Score.
- Social Pillar Score.
- Corporate Governance Pillar Score.
- ESG Controversies Score:

Controls:

- Non-Performing Loan Ratio (NPLR).
- Capital Adequacy (cap_adq).
- Profitability (prof).
- Liquidity (liq).
- Debt-to-Equity Ratio (DtE).
- Loan-to-Deposit Ratio (LtD).
- Bank Size (log asset).
- GDP growth (gdp_gr).
- Country dummies (for country-level fixed effects).

02.

Title: ESG as a Measure of Credit Ratings.

Authors: Patrycja Chodnicka-Jaworska.

Published: 2021 (Risks, 9(12), 226).

URL: https://www.mdpi.com/2227-9091/9/12/226.

Summary: The study analyses the impact of ESG measures on the long-term debt ratings assigned by Moody's and Fitch to European non-financial corporates, considering sectoral differences. Using ordinal logit models on panel data, the author tests whether and how the individual ESG pillars (environmental, social, governance, and controversies) influence credit ratings. The study highlights that ESG factors, especially environmental ones, have become relevant determinants in ratings revisions, especially during and after the Covid-19 pandemic.

Theoretical framework: The work is based on the moral capital theory (Godfrey, 2005) and on the insurance-like effect hypothesis of ESG practices, according to which such practices strengthen corporate resilience and reduce perceived risk in contexts of uncertainty or crisis. Furthermore, it refers to the stakeholder theory, according to which good ESG management strengthens stakeholder trust and, indirectly, creditworthiness.

Research Questions & Hypotheses:

Research Questions:

Do ESG measures influence Moody's and Fitch credit ratings of European non-financial companies?
Does the impact of ESG factors on ratings vary by economic sector?

Hypotheses:

H1: Changes in ESG risks negatively influence changes in credit ratings.
H2: The impact of ESG factors on credit ratings varies significantly across economic sectors.

Empirical findings (confirmation or not of the hypotheses):

H1: Confirmed. Changes in ESG scores, particularly in the environmental component and ESG controversies, are associated with significant changes in ratings.
H2: Confirmed. The effect of ESG factors is differentiated by sector: for example, the environmental score is more relevant in the energy, materials, and utilities sectors, while the social score has a greater impact in the healthcare and consumer non-cyclicals sectors.

Research Method: Ordinal logit model for panel data. Ratings are treated as ordinal variables, converted to a numerical scale by linear decomposition.

Period: 2010–2020 (with retrospective data from 1990 and predominant use of the decade 2010–2020 for quality and availability of ESG data).

Sample & Data:

9521 listed European non-financial companies.
Long-term debt ratings issued by Moody's and Fitch.
ESG data from Refinitiv (Thomson Reuters), financial data from Bankscope, and World Bank. Quarterly frequency.

Variables & Controls:

Dependent:

- Long-term credit rating (issuer rating) from Fitch and Moody's, converted into a numeric variable (scale 0–100).

Explanatory:

- ESG Combined Score (ESGC).
- ESG Score (self-reported).
- Environmental Pillar Score (ENV).
- Social Pillar Score (SOC).
- Corporate Governance Pillar Score (GOV).
- ESG Controversies Score (CONT).

Controls:

- Profitability indicators (Gross Margin, Pretax ROA).
- Solidity indicators (Debt/Equity, Long-term debt/capital, EBITDA coverage).
- Liquidity indicators (Current Ratio, Time Interest Earned).
- Operational efficiency indicators (Asset Turnover, Inventory Turnover, A/R Turnover).
- Reinvestment and growth indicators (Earnings Retention, Reinvestment Rate, WC/Sales Growth).
- Sector dummies and control for company size.

6.3 ESG and Cost of Capital: Quantitative Analyses and Debates

The relationship between ESG performance and the cost of capital of non-financial companies is a topic of growing attention in academic literature and financial practice. Integrating ESG criteria into corporate strategies is no longer just an ethical choice; it is also increasingly being assessed in terms of its economic and financial impact. In particular, a solid ESG performance can favourably influence the cost of capital, understood as both the cost of debt and the cost of equity, reducing the perceived risk by investors and financiers. Recent literature (2020–2025) has empirically analysed this relationship, generally finding a negative association between ESG scores and the cost of capital. However, the intensity and significance of the effect may vary depending on the geographical context, the industrial sector, the specific ESG pillars considered (environmental, social, or governance), and the measurement methodologies adopted. This review collects the main contributions published in journals indexed by Scopus and Web of Science, to summarise the available evidence, identifying the theoretical transmission mechanisms and outlining the managerial and policy implications. As already clarified, recent literature, focused in particular on the period 2020–2025, highlights an evolution in the recognition of ESG not only as a reputational or compliance element, but as an effective determinant of corporate financing conditions. From a theoretical point of view, the contributions analysed refer to a shared core of conceptual frameworks. Stakeholder theory (Freeman, 1994) is the most recurrent reference and interprets ESG as a lever for reducing conflicts with stakeholders, favouring more favourable conditions of access to capital. This perspective integrates with the signalling theory (Spence, 1973), according to which ESG disclosure represents a signal of management quality and lower risk, and with the information asymmetry theory (Diamond, 1989), which links ESG transparency to the reduction of risk premiums required by creditors and investors. From an empirical point of view, the results converge in recognising a negative and significant relationship between ESG and the cost of capital, albeit with important differences between its components. The study by Fiorillo et al. (2025), based on a large global sample of bond issues, shows how companies with high ESG ratings benefit from a lower return at the time of issuance. This effect is mainly attributed to environmental and social scores, while governance is less relevant. Consistently, Chen et al. (2023) demonstrate that, in the Chinese context, a high ESG score is associated with a lower cost of equity capital, thanks to the reduction of systemic risk and the expansion of the shareholder base. Confirming these mediated effects, the authors show that the improvement of ESG contributes to a reduction of the risk premium, consistent with the CAPM model. The study by Shi et al. (2024), instead, focuses on the cost of debt in the Chinese context, using the ESG scores provided by the China Securities Index. The results show that all three ESG pillars contribute to reducing the cost of debt, with governance as the main driver. However, the effect is only evident on bank loans, and not on corporate bonds, suggesting a different sensitivity between financing instruments. A relative exception comes from Saputra and Rahman (2024), who, analysing a sample of ASEAN companies, find a positive correlation between the aggregate ESG score and the cost of capital. However, the disaggregated analysis shows that the environmental score significantly reduces the cost of debt, while there is no effect on the cost of equity capital. The inconsistency may reflect differences in development in financial markets or in ESG rating systems in emerging contexts. In summary, the literature reviewed suggests that ESG is a key determinant in risk assessment and capital pricing, especially in regulated and transparent markets. Environmental and social dimensions are frequently associated with lower risk premiums, while governance shows more heterogeneous effects. Challenges remain related to the quality and homogeneity of ESG scores, transparency of disclosure, and the impact of sustainability regulations, such as the SFDR regulation in Europe.

01.

Title: Impact of the Environmental, Social, and Governance Rating on the Cost of Capital: Evidence from the S&P 500.

Authors: Dietmar Ernst, Florian Woithe.

Published: 2024 (Journal of Risk and Financial Management, 17(3), 91).

URL: https://www.mdpi.com/1911-8074/17/3/91.

Summary: The study analyses the impact of ESG scores on the cost of capital for S&P 500 companies. Using Bloomberg financial data and MSCI ESG scores for 498 companies, the authors explore the relationship between ESG scores and the three components of the cost of capital: cost of equity, cost of debt, and WACC. The results show that high ESG scores are associated with a lower cost of equity and debt, but not with a lower WACC. This is explained by a difference in capital structures: companies with lower ESG ratings have higher leverage, which artificially lowers the WACC.

Theoretical framework: The study is implicitly based on two main theoretical frameworks: (i) Financial risk and cost of capital theory (Capital Asset Pricing Model—CAPM): The basis of the analysis of the cost of equity capital is the concept according to which the systematic risk of the company (measured by beta) determines the expected return required by investors. The implicit assumption is that a high ESG rating may signal a lower exposure to systemic, reputational, environmental or regulatory risks, with a consequent reduction in beta and therefore in the cost of equity capital; (ii) Stakeholder theory: A good ESG performance reflects a careful management of stakeholders and a commitment to sustainability, factors that can strengthen the confidence of investors and financiers, improve the company reputation and, ultimately, reduce the perceived risk. This risk reduction, according to the logic of efficient markets, translates into a lower cost of capital. In summary, the theoretical logic guiding the study is that ESG performance acts as a proxy for management quality and long-term sustainability, with direct (risk reduction) and indirect (facilitated access to capital) impacts on the cost of financial resources.

Research Questions & Hypotheses:

Research Questions:

How do ESG scores affect the cost of equity capital (CAPM beta) for S&P 500 companies?
Are ESG scores associated with a lower cost of debt?
Is there a relationship between ESG score and WACC?

Hypotheses:

H1: A higher ESG score is associated with a lower beta and therefore a lower cost of equity capital.
H2: A higher ESG score is associated with a lower cost of debt.
H3: A higher ESG score is associated with a lower WACC.

Empirical findings (confirmation or not of the hypotheses):

H1: Confirmed: higher ESG scores are associated with lower beta and lower cost of equity.
H2: Confirmed: The cost of debt is lower for firms with better ESG ratings.
H3: Not confirmed: Companies with higher ESG scores have a slightly higher WACC, due to their lower leverage compared to "Laggards".

Research Method: (i) Descriptive analysis, Pearson correlations and linear regressions (with normality and homoscedasticity transformations); (ii) Beta, costs of capital and WACC obtained via Bloomberg. ESG score from MSCI.

Period: Data collected on a single date: November 24, 2022.

Sample & Data:

498 companies in the S&P 500 index.
ESG rating (AAA–CCC) and scores (0–10) from MSCI.
Financial data from Bloomberg (cost of debt, cost of equity, WACC, beta).

Variables & Controls:

Dependent:

- Beta (proxy of the cost of equity capital).
- Cost of debt.
- WACC

Explanatory:

- Overall ESG Score from MSCI.

Controls:

- Not specified in detail in the base models, but the paper includes considerations of leverage as a variable affecting WACC.

Others:

- The effect of the ESG score on the WACC is not direct but mediated by the capital structure: ESG-compliant companies have higher equity and lower debt, resulting in an increase in the WACC despite the lower individual cost of the sources.

02.

Title: ESG Performance and Its Impact on Mitigating Cost of Capital: Evidence from Southeast Asia.

Authors: Anggi Saputra, Annisaa Rahman.

Published: 2024 (E-Jurnal Akuntansi, 34(8), 2054–2072).

URL: https://ojs.unud.ac.id/index.php/akuntansi/article/view/116855.

Summary: The study analyses the impact of ESG performance on the cost of capital in non-financial firms in five developing countries in Southeast Asia (Indonesia, Vietnam, Thailand, Malaysia, and the Philippines), over the period 2018–2023. The results highlight a positive correlation between the aggregate ESG score and the cost of capital, with a significant effect of the environmental component only. However, the disaggregated analysis shows that ESG reduces the cost of debt but has no effect on the cost of equity.

Theoretical framework: The theoretical framework is based on: (i) Signalling theory (Spence, 1973): the disclosure of ESG information acts as a positive signal towards stakeholders, improving the perception of risk and the expected value of the company; (ii) Legitimacy theory (Dowling & Pfeffer, 1975): ESG practices strengthen the social legitimacy of the company, improving its reputation and access to capital; (iii) The concept of sustainability as a lever for reducing perceived risk is recalled (Sharfman & Fernando, 2008), with direct implications on the cost of capital.

Research Questions & Hypotheses:

Research Questions:

Does ESG performance influence the cost of capital in Southeast Asian firms?
To what extent do the three ESG pillars (E, S, G) impact the cost of capital?
Are there distinct effects on the cost of debt and the cost of equity?

Hypotheses:

H1: ESG performance harms the cost of capital.

H2: The three ESG pillars negatively influence the cost of capital.
H3: ESG reduces both the cost of debt and the cost of equity.

Empirical findings (confirmation or not of the hypotheses):

H1: Hypothesis 1 is refuted: aggregate ESG shows a positive correlation with the cost of capital.
H2: Hypothesis 2 is partially confirmed: only the Environmental Score has a significant impact.
H3: Hypothesis 3 is partially confirmed: ESG reduces the cost of debt, but not that of equity.

Research Method: (i) Panel data regression (STATA 17); (ii) Disaggregated analysis of the three ESG pillars.
Period: 2018–2023.
Sample & Data:

Non-financial corporations listed on stock exchanges in ASEAN developing countries.
Sources: corporate financial statements, ESG ratings from local/regional sources.

Variables & Controls:

Dependent:

- Cost of Capital (CoC), Cost of Debt, Cost of Equity.

Explanatory:

- Aggregate ESG Score.
- Environmental Score.
- Social Score.
- Governance Score.

Controls:

- Company size, leverage, liquidity, profitability, and macro variables.

Others:

- Separate analysis for CoC, debt, and equity.

03.
Title: ESG Performance and the Cost of Debt. Evidence from the Corporate Bond Market.
Authors: Paolo Fiorillo, Antonio Meles, Antonio Ricciardi, Vincenzo Verdoliva.
Published: 2025 (International Review of Financial Analysis, 102, 104097).
URL: https://www.sciencedirect.com/science/article/pii/S105752192500184X?via%3Dihub.
Summary: The study analyses the relationship between ESG performance and the cost of debt in the primary corporate bond market. Based on a large international sample of 25,234 bond issues by 2677 ESG-rated firms, the authors show that firms

with high ESG ratings issue bonds with a lower yield of about 10–11 basis points compared to those with low ESG ratings. The effect is mainly attributed to the environmental and social dimensions, while governance has no relevant impact. The effect of reducing the cost of debt is particularly significant in countries with developed financial markets and more effective regulations (e.g., creditor protection). Furthermore, a positive effect resulting from the introduction of the European SFDR regulation in 2021 is found.

Theoretical framework: The theoretical framework is based on three main pillars: (i) Market efficiency theory: in the most developed and efficient financial markets, ESG information is more easily incorporated into bond prices, with direct impacts on the expected return by subscribers; (ii) Stakeholder theory: a better ESG performance signals responsible and long-term oriented management, capable of attracting investors and reducing perceived risk; (iii) Legitimacy theory: alignment with environmental and social sustainability objectives increases the legitimacy of the company in the market, facilitating access to capital at more favourable conditions.

Research Questions & Hypotheses:

Research Questions:

- Is there an ESG premium (negative spread) in the primary corporate bond market?
- What is the effect of individual ESG pillars (E, S, G) on the yield spread?
- Has the introduction of the SFDR regulation in Europe influenced the ESG premium?

Hypotheses:

- H1: Companies with high ESG ratings enjoy lower subscription returns.
- H2: ESG does not influence performance or has an inverse effect.

Empirical findings (confirmation or not of the hypotheses):

- H1: Confirmed. The average spread between high and low ESG issuers is −10/−11 bps.
- H2: Not confirmed. No systematic evidence is found to support either no or positive effects.

The effect is driven by environmental and social dimensions, while governance has no significant impact. The ESG premium is more pronounced post-SFDR (2021) and in more advanced institutional contexts.

Research Method: (i) Multivariate regressions with endogeneity control techniques (IV tools, propensity score matching, robustness analysis); (ii) Matching between similar firms (high ESG vs low ESG) on company and bond characteristics.

Period: 2002–2024 (with analytical focus on the period 2018–2024 and pre/post SFDR comparison).

Sample & Data:

25,234 bond issues.

2677 companies with ESG ratings.
Data from Refinitiv, Bloomberg, Moody's, Fitch, R&I, and other ratings providers.
International coverage (non-financial companies).

Variables & Controls:

Dependent:

- Yield to Maturity (YTM) at the time of bond issuance.

Explanatory:

- Aggregate ESG Rating.
- Environmental Score.
- Social Score.
- Governance Score.

Controls:

- Bond duration, seniority, currency, size of issue, sector, credit rating level, characteristics of the issuing company.
- Regulation (pre/post SFDR).
- Size and development of the national financial market.
- Quality of the bankruptcy legal framework.

Others:

- The green/social labelling of the bond does not explain the ESG premium: what matters is the ESG performance of the issuer.
- The ESG effect is stronger in regulated and transparent contexts, suggesting a key role of institutions.

04.
Title: Effect of ESG Performance on the Cost of Equity Capital: Evidence from China.

Authors: Yonghuai Chen, Tao Li, Qing Zeng, Bo Zhu.

Published: 2023 (International Review of Economics and Finance, 83, 348–364).

URL: https://www.sciencedirect.com/science/article/abs/pii/S1059056022002258?via%3Dihub.

Summary: The study investigates the effect of ESG performance on the Cost Of Equity (COE) of Chinese A-share listed companies between 2010 and 2020. Using data from the Sino-Securities ESG Index, the authors demonstrate that a better ESG score significantly reduces COE, both directly and indirectly, by reducing market risk and increasing equity diversification. The results are robust to heteroskedasticity, autocorrelation, and cross-correlation issues.

Theoretical framework: The theoretical framework is based on two main approaches: (i) Risk mitigation theory: ESG is considered a mechanism capable of reducing the systemic and specific risk of the company, decreasing the risk premium required by investors (Cheng et al., 2014; Dimson et al., 2015); (ii) Stakeholder theory: ESG performance improves the relationship with key stakeholders, helping to stabilize cash flows and reduce earnings volatility, with a positive impact on risk assessment and the cost of capital (Freeman, 1994; Cek & Eyupoglu, 2020). The sustainable growth model and the Kuznets environmental curve are also recalled to support the link between sustainable development and long-term competitiveness.

Research Questions & Hypotheses:

Research Questions:

Does ESG performance affect the cost of equity of Chinese listed firms?
What are the transmission mechanisms of the ESG effect on the COE?
Do the three ESG pillars have different impacts?

Hypotheses:

H1a: ESG performance is positively related to COE (cost effect).
H1b: ESG performance is negatively related to COE (benefit effect).
H2: ESG reduces COE indirectly through lower market risk and greater equity diversification.

Empirical findings (confirmation or not of the hypotheses):

H1a: Not confirmed.
H1b: Confirmed. A higher ESG score is associated with lower COE.
H2: Confirmed. Evidence of significant mediation effects through risk reduction and equity diversification.

Research Method: (i) Two-way fixed effects model for panel data; (ii) Mediation analysis (mediation model); (iii) Robustness test with different ESG data sources (Sino-Securities, Wind, Bloomberg).

Period: 2010–2020.

Sample & Data:

Over 1500 companies are listed on China's A-share market.
ESG scores from Sino-Securities Index (core), Wind, and Bloomberg (robustness).
COE is estimated with the Capital Asset Pricing Model (CAPM).
Accounting and financial data from CSMAR and Wind.

Variables & Controls:

Dependent:

- Cost of Equity Capital (re), estimated with CAPM.

Explanatory:

- Aggregate ESG score (Sino-Securities).
- ESG disaggregated dimensions: E, S, G.

Controls:

- Company size, sales growth, leverage, book-to-market, tax shield, turnover, nature of ownership.
- Sector and year effects.

Others:

- Mediating variables: systemic market risk, degree of stock diversification.

05.

Title: ESG Performance and Cost of Debt.

Authors: Yongdong Shi, Shijie Zheng, Pengsong Xiao, Hongxian Zhen, Tong Wu.

Published: 2024 (China Journal of Accounting Research, 17, 100390).

URL: https://www.sciencedirect.com/science/article/pii/S1755309124000480?via%3Dihub.

Summary: The paper examines the impact of ESG performance on the cost of debt in Chinese firms, using the ESG score developed by China Securities Index Co., Ltd. (CSI), which integrates local specificities and international criteria. The results highlight that higher ESG scores are associated with a lower cost of debt, especially for bank loans. Among the three ESG pillars, governance has the strongest effect on reducing the cost of debt, followed by social, and finally environmental. The effect is amplified by the quality of local environmental execution and attenuated by the quality of internal controls and the degree of marketisation.

Theoretical framework: The theoretical framework is built on a comparison between two opposing perspectives: (i) Shareholder expense view/Agency cost theory: the adoption of ESG practices generates additional costs that reduce returns for shareholders and potentially increase risk for creditors (Friedman, 2007; Krüger, 2015) → Foundation of H1-a; (ii) Stakeholder theory/Risk mitigation theory: a solid ESG performance strengthens corporate reputation, improves the relationship with stakeholders, reduces information asymmetry and therefore lowers the risk premium required by creditors (Freeman, 1994; Cheng et al., 2014) → Foundation of H1-b.

Research Questions & Hypotheses:

Research Questions:

Does the ESG score, influenced by domestic and international standards, have a significant impact on the cost of debt for Chinese enterprises?

Which of the three ESG pillars has the strongest effect?

What are the mechanisms through which ESG influences the cost of debt?

Hypotheses:

H1-a: ESG performance is positively related to the cost of debt → ESG is a non-productive cost factor for creditors.

H1-b. ESG performance is negatively related to the cost of debt → ESG is a tool for reducing risk and the cost of financing.

Empirical findings (confirmation or not of the hypotheses):

H1-b: Confirmed. The analysis shows that firms with higher ESG scores have a lower cost of debt, especially for bank loans.

H1-a: Not confirmed. There is no positive effect of ESG on the cost of debt. Among the three ESG pillars, governance has the largest effect, followed by social and environmental. The negative effect of ESG on the cost of debt is stronger in contexts with high environmental enforcement and a low level of marketisation.

Research Method: (i) Panel regressions with controls for endogeneity (IV instruments); (ii) Disaggregated analysis by source of debt: bank loans vs. bonds; (iii) Analysis of transmission mechanisms (mediated by risk and information asymmetry).

Period: 2018–2020.

Sample & Data:

Companies listed in the main CSI 300 and CSI 500 indices.
ESG score from CSI (China Securities Index).
Data on financing (loans and bonds) from CSMAR and WIND.

Variables & Controls:

Dependent:

- Cost of Debt (weighted average cost of bank debt).

Explanatory:

- Overall CSI ESG Score.
- Environmental, Social, Governance scores (disaggregated).

Controls:

- ROA, leverage, size, tax shield, quality of internal controls, regional and sector dummies.
- Institutional indicators: environmental enforcement, marketisation index.

Others:

- ESG reduces the cost of debt only for bank loans, not for bonds.
- The ESG-debt link is mediated by a better perceived repayment capacity and by lower information asymmetry.
- The governance component is crucial in improving financiers' confidence.

6.4 ESG and Firm Value: Methodologies and Outcomes

The relationship between ESG indicators and enterprise value is a topic of growing relevance in corporate finance studies and managerial practice. The intensification of pressure from stakeholders and investors for responsible management in terms of environmental, social, and governance (ESG) has stimulated companies to incorporate these criteria into their operational and financial strategies. The aim is to create sustainable value in the long term, while increasing attractiveness towards the financial market and improving reputational perception. This review explores the most recent academic studies, published between 2020 and 2025, that analyse the influence of ESG performance on corporate value, measured through financial indicators such as market value, profitability, and the persistence of value itself. The selected articles mainly use quantitative methodologies based on linear and non-linear regression models, moderation, and survival analyses, providing empirical results that confirm, in the majority of cases, a positive relationship between high ESG ratings and corporate value. However, there is also evidence of non-linear or specific relationships to individual ESG components, indicating that the impact is not homogeneous and may vary across geographic contexts, industry sectors, and company characteristics. Overall, recent literature highlights the growing importance of ESG aspects as strategic factors capable of significantly influencing corporate valuation in contemporary financial markets.

01.

Title: The Impact of ESG on Firm Risk and Financial Performance: A Systematic Literature Review.

Authors: Hua Nian, Fathin Faizah Said.

Published: 2024 (Journal of Scientometric Research, 13(3s), pp. s144–s155).

URL: https://jscires.org/article/7878/.

Summary: The article proposes a systematic literature review on the impact of ESG indicators on corporate risk and financial performance, focusing on studies published between January 2013 and September 2023 in the Scopus and Web of Science databases. The aim is twofold: to map the main methodological trends and to highlight the prevailing conclusions. 35 articles on the ESG-Risk relationship and 120 on the ESG-Financial Performance relationship were analysed. The study also focuses on regional specificities, with a particular focus on East Asia and ASEAN countries.

Theoretical framework: Although the work is a systematic review of literature, the underlying theoretical bases emerge implicitly from the references to the main theoretical strands adopted in the studies analysed. Among the most recurrent references are: (i) Stakeholder Theory: companies that respond to the interests of a plurality of stakeholders—and not only shareholders—can improve their competitive positioning and value creation in the long term, reducing reputational and operational risks; (ii) Legitimacy Theory: ESG practices are considered a means to maintain and strengthen the social legitimacy of the company, an essential condition for the sustainability of the activity over time; (iii) Resource-Based View (RBV): ESG management capabilities are seen as distinctive resources that confer sustainable competitive advantages, contributing to improving performance and containing risks; (iv) Risk Management Perspective: the integration of ESG factors is interpreted as a tool to anticipate and mitigate financial, regulatory and environmental risks, with potential positive effects on the stability of corporate value. These theoretical approaches provide an interpretive framework for the different ways in which ESG performance influences corporate risk and performance.

Research Questions & Hypotheses:

Research Questions:

What are the trends and characteristics of research on the ESG-business risk and ESG-financial performance relationship?
What methods are used to study this relationship?
What effects emerge in the literature?
What are the main research gaps?

Hypotheses:

The article does not formulate explicit hypotheses, as it is a systematic review. However, it aims to verify the presence of a consensus in the literature on the positive, negative, neutral or curvilinear impact of ESG on risk and financial performance.

Empirical findings (confirmation or not of the hypotheses):

ESG and business risk:

68.57% of articles claim that ESG reduces corporate risk. 17.14% report mixed effects (depending on the type of risk). 8.6% find no significant effects. 5.7% observe an increase in risk. No articles show curvilinear relationships.

ESG and financial performance:

50% of studies show positive effects. 13.33% find negative effects. 20.83% report mixed effects. 8.33% find no significant impact. 7.5% find curvilinear relationships.

Research Method: Systematic review of the literature according to PRISMA protocol. The selected articles mainly use fixed effects models on panel data, followed by pooled models, PVAR, quantile regressions, dynamic GMM, SEM, and others. The methods are more varied in the studies on ESG-Performance than in those on ESG-Risk.

Period: Articles published between January 2013 and September 2023.

Sample & Data:

35 articles on ESG and business risk.
120 articles on ESG and financial performance.
All from Scopus and WoS. Research on East Asia and ASEAN countries has emerged more recently but is growing rapidly.

Variables & Controls:

Dependent:

- Business risk (idiosyncratic, systemic, credit, stock market crash, among others).
- Financial performance (ROA, ROE, ROI, Tobin's Q, stock return, and EPS).

Explanatory:

- Aggregate ESG indicators (rarely the individual E, S, G components).

Controls:

- Company and sector variables (size, geographical area, sector, economic cycle, among others).

Others:

- Classification of research methods and analysis of regional differences.
- Considerations on issues of heterogeneity and standardisation of ESG ratings.

02.

Title: The Impact of ESG on Firm Value: The Moderating Role of Cash Holdings.

Authors: Omnia Ahmed, Bashar Abu Khalaf.

Published: 2025 (Heliyon, 11(e41868)).

URL: https://www.sciencedirect.com/science/article/pii/S2405844025002488?utm_source=chatgpt.com.

Summary: The article examines the relationship between ESG performance and the market value of European companies, introducing cash holdings as a moderating variable. The analysis is based on a sample of 1144 non-financial companies operating in the 27 countries of the European Union between 2013 and 2023. The authors find that a higher ESG performance is significantly associated with a higher market value, and that the presence of significant cash reserves strengthens this relationship. In other words, companies that adopt ESG practices and hold adequate cash resources obtain a higher valuation by the market, as they are perceived as more resilient, sustainable, and capable of investing in strategic ESG initiatives. Furthermore, cash reserves improve the capacity to absorb external shocks and support sustainability-oriented investments, especially in the regulated contexts of the European Union.

Theoretical framework: The theoretical framework of the article is based on a complex combination of four main theories: (i) Resource Dependence Theory (RDT): Companies depend on external resources and must be able to manage them effectively to survive and prosper. The adoption of ESG practices allows for to reduction of dependence on critical resources, improves the management of stakeholder relationships, and promotes adaptation to the external context. Liquid reserves represent a strategic asset that allows companies to implement ESG policies flexibly and proactively; (ii) Agency Theory (Lee & Isa, 2022): There are conflicts between managers and shareholders on the use of company resources. ESG initiatives can be used by managers to pursue personal interests (e.g., improving the image), with the risk of window dressing. However, effective governance and control over liquid reserves can mitigate agency conflicts and orient ESG choices towards the creation of value for shareholders; (iii) Stakeholder Theory: Companies must respond not only to shareholders, but to all stakeholders. Liquid resources allow companies to invest in ESG programs that respond to the interests of workers, communities, customers and suppliers, helping to improve reputation, reduce risks and strengthen trust in the long term; (iv) Stewardship Theory (Katelouzou & Klettner, 2022): Managers, if motivated and well structured, can act as stewards of the collective interests of the company, using available resources—including liquid reserves—to carry out authentic ESG projects consistent with a sustainable vision. In this sense, the theory emphasises the ability of managers to strengthen corporate value through ethical and responsible decisions.

Research Questions & Hypotheses:

Research Questions:

To what extent does ESG performance affect the market value of companies?
Do liquid assets moderate this relationship?

Hypotheses:

H1: ESG ratings are positively associated with firms' market value.
H2: Liquid assets reinforce the positive effect of ESG performance on market value.

Empirical findings (confirmation or not of the hypotheses):

H1 confirmed. There is a positive and significant relationship between ESG (and each component: E, S, G) and enterprise value.
H2 confirmed. The interaction between ESG and cash holdings is significant; cash reserves amplify the positive effect of ESG on firm value.

The other control variables (dividends, R&D, ROA) are also positively associated with firm value.

Research Method: The empirical analysis is based on a panel regression model (fixed and random effects), applied to 1144 European companies in the period 2013–2023, with data from LSEG Data & Analytics. After the Hausman test, the fixed effects model was adopted. To support the robustness of the results, the authors also apply a Tobit regression, which confirms the significance of the identified relationships.

Period: 2013–2023.

Sample & Data:

1144 non-financial companies from the 27 EU countries.
Excluding financial sectors and companies with more than 4 years of missing data.
Sources: LSEG Data & Analytics.

Variables & Controls:

Dependent:

- Market Value (MV).

Explanatory:

- ESG Score (and individual components ES, SS, GS).
- ESG × Cash Holdings (interaction).

Controls:

- Company and sector variables (size, geographical area, sector, economic cycle, among others).

Others:

- Cash Holdings (CH).
- Dividends (DIV).
- Research & Development (RD).
- Profitability (ROA).

03.

Title: Valuing ESG: How Financial Markets Respond to Corporate Sustainability.

Authors: Le Chau, Le Anh, Vo Duc.

Published: 2025 (International Business Review, 34(3), 102418).

URL: https://www.sciencedirect.com/science/article/pii/S0969593125000319?utm_source=chatgpt.com.

Summary: The paper investigates the relationship between ESG performance and firm value, assuming a non-linear and cubic relationship. The authors argue that the stock market does not react uniformly to different levels of ESG commitment: while an initial improvement in ESG practices can increase firm value, an intermediate phase characterised by costly investments and uncertain outcomes can generate a negative impact, followed by a new increase in value for the most advanced and resilient firms in terms of sustainability. The study uses a large dataset of companies listed in 12 countries between 2015 and 2021, combining ESG data (Refinitiv) with financial and institutional variables. The notion of S-shape relationship is introduced, where the cubic behaviour reflects the underlying dynamics between the growth of ESG capabilities, stakeholder management, and market response. The results show that this relationship is particularly pronounced in emerging markets with low institutional quality, where ESG has a stronger signalling function.

Theoretical framework: The theoretical framework is structured on several levels: (i) Real Options Theory: ESG practices are interpreted as real options that offer strategic flexibility. In the initial phase, the adoption of ESG allows for seizing reputational opportunities at low cost. In the intermediate phase, costs increase in the face of uncertain benefits. Finally, in the advanced phase, ESG-mature companies capitalise on long-term benefits; (ii) Stakeholder Theory and SIC: The effectiveness of ESG practices depends on the ability of companies to generate value through solid relationships with stakeholders; (iii) Institutional Theory: The national institutional context influences the relevance of the ESG signal. In advanced countries, ESG is now a standard; in countries with weak institutions, it represents a powerful signal of reliability; (iv) Agency Theory: An intermediate phase characterised by an excessive focus on social issues can fuel agency conflicts.

Research Questions & Hypotheses:

Research Questions:

How does the market value of the company respond to different levels of ESG performance?
What is the role of the institutional context in the ESG–corporate value relationship?

Hypotheses:

H1: The relationship between ESG ratings and firm value follows a cubic pattern, where value initially increases as ESG ratings increase up to a critical threshold, then decreases, and finally increases again above a second threshold.
H2: The effect of ESG performance on firm value is stronger in countries with weaker sustainability norms and regulatory systems.
H3: The relationship between ESG and firm value is moderated by the institutional context, through the interaction between ESG performance and national sustainability indicators.

Empirical findings (confirmation or not of the hypotheses):

H1: Confirmed. The cubic regression shows the expected trend: positive in the early stages, negative in the middle stage (when ESG costs exceed benefits), and positive again in the late stage.
H2 and H3: Confirmed. The ESG effect on value is greater in weak institutional contexts, where ESG acts as a signal of quality. The interaction between ESG and national sustainability indicators is statistically significant.

Research Method: The study uses a cubic polynomial regression on panel data, including ESG^2 and ESG^3 terms, to model the non-linear behaviour of the ESG-value relationship. The dependent variable is the Tobin's Q, while the main explanatory variable is the ESG score (Refinitiv), with additional company control variables and interactions with institutional indicators. Among the main techniques used by the authors, there is panel regression with controls for industry, country, and year. Robustness verified through lagged ESG variables, GMM (Generalized Method of Moments), analysis of regional subsamples (Europe, Asia, among others), and instrumental variables.

Period: 2015–2021.

Sample & Data:

Over 4000 listed companies in 43 countries.
ESG Source: Refinitiv.
Institutional quality: World Bank Governance Indicators + national regulatory ESG indicators.

Variables & Controls:

Dependent:

- Tobin's Q (market value to book value).

Explanatory:

- ESG score.
- ESG^2.
- ESG^3.
- ESG × institutional quality.

Controls:

- Size of the company.
- ROA.
- Leverage.
- R&D intensity.
- Country, sector, and year fixed effects.

04.
Title: Beyond compliance: evaluating the role of environmental, social, and governance disclosures in enhancing firm value and performance.

Authors: Phemelo Tamasiga, Helen Onyeaka, Malebogo Bakwena, El houssin Ouassou.

Published: 2024 (SN Business & Economics, 4(118)).

URL: https://link.springer.com/article/10.1007/s43546-024-00714-6?utm_source=chatgpt.com.

Summary: The article presents a systematic review of the literature aimed at exploring the role of ESG disclosures in the relationship between ESG performance, firm value, and profitability. While the literature has already widely discussed the direct link between ESG performance and financial performance, the originality of the contribution lies in considering the mediating role of ESG disclosures. The analysis is based on 52 studies selected from Scopus and Web of Science, relating to companies operating in numerous sectors and geographical areas. The results highlight that the quality and transparency of ESG disclosures can significantly influence stakeholders' perception and, consequently, the market value and profitability of the company. The implications vary according to the sector and the institutional context. In the food and retail sectors, a positive correlation between ESG disclosure and financial performance is observed. In developing countries, however, the relationship is often null or negative. The role of governance disclosures in improving the market valuation of agricultural companies is particularly relevant.

Theoretical framework: The theoretical framework of the article is based on a combination of approaches that highlight the relevance of ESG disclosures through and signalling: (i) Stakeholder Theory: ESG disclosures are tools through which companies communicate their commitment to sustainability, strengthening trust and legitimacy among stakeholders; (ii) Signalling Theory: Companies use ESG disclosures to send signals of quality and responsibility to the markets. This mechanism is crucial, especially in emerging markets; (iii) Institutional Theory: The regulatory and institutional context modulates the quality of disclosures and their effectiveness in generating trust; (iv) Legitimacy Theory: Disclosures contribute to maintaining the social and institutional legitimacy of the company.

Research Questions:

Is there a quantifiable relationship between ESG scores and corporate profitability, considering the mediating effect of disclosures?
How do the quality and transparency of ESG disclosures impact stakeholder trust?
To what extent do regulatory frameworks influence the quality of ESG disclosures and their impact on corporate value and performance?

Hypotheses:

It should be noted that no explicit hypotheses are formulated, as this is a systematic review of the literature, but the research questions reflect the intent to identify a mediating role of ESG disclosures between ESG scores and economic-financial performance.

Empirical findings (confirmation or not of the hypotheses):

Mediating role: ESG disclosures strengthen the positive relationship between ESG performance and corporate value/performance, especially in the food, retail, and utility sectors.
Sectorality: In regulated sectors (e.g., utilities), social aspects prevail over environmental ones in disclosures; in agriculture, governance disclosures improve market valuation.
Geographic context: In developing countries, the ESG-performance relationship is weak or negative.
Divergent effects: Environmental disclosure in some sectors does not generate positive impacts, highlighting the need for contextualised and non-standardised approaches.

Research Method: The study is conducted through a rigorous systematic literature review (SLR), according to the Prisma 2020 protocol, aimed at ensuring transparency, reproducibility, and methodological robustness. Main steps: (i) Databases used: Scopus and Web of Science; (ii) Extraction date: December 15, 2023; (iii) Tools: use of Biblioshiny software (R interface) for descriptive analysis of bibliometric data; (iv) Inclusion criteria: peer-reviewed articles in English, focused on the link between ESG disclosure and corporate performance (value and profitability); (v) Screening: out of 133 initial documents, 52 articles selected after removal of duplicates and full-text screening. The study integrates quantitative and qualitative evidence on a cross-sector and international scale, including studies conducted in developed, emerging, and developing contexts.

Period: Articles published up to December 2023 (mainly concentrated from 2015 to 2023).

Sample & Data:

52 studies selected from Scopus and WoS.

Geographic scope: Global (including OECD countries, the MENA region, Asia, and Latin America).
Sectors analysed: banking, utilities, retail, manufacturing, agriculture, energy, telecommunications.

Variables & Controls:

Dependent:

- Market value (Tobin's Q, market-to-book).
- Profitability (ROCE, ROA, net margins).

Explanatory:

- ESG scores (aggregated or disaggregated).
- ESG disclosure quality.
- ESG materiality.

Controls:

- Company age, size, sector, regulatory context, and institutional development.

05.
Title: The Effect of Environmental, Social, and Governance (ESG) on the Persistence of Firm Value: Evidence from Survival Analysis.
Authors: Yen-Yu Liu, Pin-Sheng Lee.
Published: 2025 (Accounting and Auditing, 1(4)).
URL: https://www.mdpi.com/3042-6618/1/1/4.

Summary: This study investigates the effect of ESG performance on firm value persistence (not just static value) over time. Using data from Taiwanese listed firms between 2016 and 2023, the authors employ a survival analysis to estimate how long firms can maintain increasing levels of market value. The main finding is that the aggregate ESG score is not significantly related to value persistence, while the S (social) score shows a positive and significant relationship. Among the sub-dimensions of the social component, only product quality and safety are significantly associated with higher value persistence. This indicates that not all ESG practices influence a firm's ability to maintain value over time in the same way, and that only visible and tangible social initiatives are rewarded by stakeholders.

Theoretical framework: The theoretical framework is based on a combination of approaches: (i) Stakeholder Theory: ESG improves trust and relationships with stakeholders, fostering stability and continuity of corporate value; (ii) Resource-Based View: ESG capabilities are considered strategic resources that improve the resilience and competitiveness of the firm; (iii) Agency Theory: ESG can be instrumentalised by managers for reputational purposes, reducing the effectiveness of initiatives and their impact on long-term performance. The distinctive contribution of the article lies in the introduction of a temporal perspective (duration) in the ESG-corporate value relationship, neglected by most of the previous literature.

Research Questions:

Is ESG performance associated with greater persistence of enterprise value over time?

Hypotheses:

H1: ESG performance is positively associated with firm value persistence.

Empirical findings (confirmation or not of the hypotheses):

The overall ESG score is not significant (hazard ratio = 0.997; p = 0.444). The S (social) score is significant (hazard ratio = 0.928; p = 0.017), especially in the subdomains: Product quality and safety (hazard ratio = 0.948; p = 0.050). The E, G scores do not show significant effects. The significant control variables are: (i) EPS: better performance → greater persistence; (ii) Sentiment: investor confidence → greater persistence; (iii) Size: company size → greater persistence.

Research Method: The study uses Cox regression with time-dependent covariates, which allows for modelling the risk of event (reduction of company value) over time, taking into account endogeneity and right-censoring. Main method: Survival analysis (Cox model with time-dependent covariates). Additional models: (i) ESG disaggregated into E, S, G; (ii) S in turn divided into: human rights, data security, product quality, employee disclosure, job security; (iii) Robustness test with lagged variables and percentile ranking.

Period: 2016–2023.

Sample & Data:

1637 listed enterprises in Taiwan.
ESG data source: Taiwan Economic Journal (TEJ), SASB compliant.
Total observations: 4267 firm-years.
Industry: technology, consumer goods, infrastructure, healthcare, energy, and transportation.

Variables & Controls:

Dependent:

- Persistence of value (survival over time of Tobin's Q).

Explanatory:

- ESG score (and E, S, G components).
- Subcomponents of the S-score.

Controls:

- EPS.
- Investor sentiment.
- COVID-19.
- Firm size (total active logs).

06.

Title: ESG and Firm Value.

Authors: Andrea Quintiliani.

Published: 2022 (Accounting and Finance Research, 11(4)).

URL: https://www.sciedu.ca/journal/index.php/afr/article/view/22864.

Summary: The paper examines the correlation between ESG score and firm value, measured both through economic-financial performance indicators (ROE, Levered Free Cash Flow, Current Ratio, Quick Ratio) and through stock price. The author employs a quantitative approach through linear regression on panel data, analysing a sample of 115 listed companies in Europe (France, Germany, Italy) in the period 2016–2020. The results show a positive and significant correlation between ESG score and each performance variable, suggesting that effective ESG practices can improve profitability, financial stability, and market valuation of the company. The paper fits into the academic debate that considers ESG not only as a compliance tool but as a strategic lever for corporate value.

Theoretical framework: The theoretical framework is mainly based on two perspectives: (i) Agency Theory: ESG practices reduce information asymmetry between managers and investors, contributing to the control of opportunistic behaviours and improving allocative efficiency; (ii) Signalling Theory: The adoption of ESG metrics acts as a quality signal for the market, positively influencing the perception of risk and the valuation of the company. Furthermore, the author recalls previous evidence linking ESG to a lower cost of capital, reinforcing the idea that sustainability acts as a determinant of the capital structure.

Research Questions:

Is there a positive correlation between financial performance and ESG score?
Does the company's market price reflect ESG performance?

Hypotheses:

H1: There is a positive correlation between corporate performance (ROE, LFCF, CR, QR) and ESG score.
H2: There is a positive correlation between market performance (Stock Price) and ESG score.

Empirical findings (confirmation or not of the hypotheses):

Both hypotheses (H1, H2) are confirmed: all independent variables (ROE, LFCF, CR, QR, Stock Price) show a positive and statistically significant relationship with the ESG score. The control variables (company size, leverage) are also significant in several models. The Adjusted R^2 is high (up to 0.62), indicating a good explanatory capacity of the models.

Research Method: Descriptive and correlational research. Multiple linear regression on panel data, with ESG as the dependent variable. Software: SPSS 20.0.

Period: 2016–2020.

Sample & Data:

115 European companies listed on Euronext Paris, Borsa Italiana, and the Frankfurt Stock Exchange
Total observations: 1225 firm-years.
ESG Source: Bloomberg ESG Disclosure Scores.

Variables & Controls:

Dependent:

- ESG Score (Bloomberg).

Explanatory:

- ROE.
- Levered Free Cash Flow.
- Current Ratio.
- Quick Ratio.
- Stock Price.

Controls:

- Leverage (Debt/Total Assets).
- Size (log assets).

6.5 ESG and Financial Leverage: Trends in Capital Structure Research

The relationship between ESG indicators and corporate capital structure is a topic of growing relevance in the corporate finance literature. The increased attention towards sustainability and the pressure exerted by institutional investors, regulators, and stakeholders has pushed non-financial firms to consider environmental, social, and governance dimensions not only as reputational elements, but also as determining factors in capital choices. In this context, decisions relating to capital structure—i.e. the optimal mix between debt and equity—are assuming a new configuration, in which ESG performance can directly influence the cost of capital, accessibility to credit, the preference for certain financial instruments (such as green bonds or ESG-linked financing) and the degree of sustainable financial leverage. This review explores the main contributions published between 2020 and 2025 to analyse the empirical and theoretical evidence relating to the ESG-Capital Structure nexus, highlighting the methodologies adopted, the prevalent results, and the explanatory mechanisms proposed in different geographical and sectoral contexts.

01.
Title: Firm ESG reputation risk and debt choice.
Authors: David P. Newton, Steven Ongena, Ru Xie, Binru Zhao.
Published: 2024 (European Financial Management, 30(4)).

URL: https://onlinelibrary.wiley.com/doi/10.1111/eufm.12468.

Summary: The paper analyses how reputational risk related to ESG practices influences the choice between bank and bond debt by US firms. The authors introduce an innovative approach using the RepRisk Index (RRI) to measure firms' exposure to negative news related to ESG issues. The central hypothesis is that a high ESG reputational risk reduces the possibility of access to bank credit, due to the greater aversion to reputational risk by banks, pushing firms towards financing through the issuance of public bonds. The analysis, conducted on over 71,000 quarterly observations of 3783 US firms between 2007 and 2020, confirms that an increase in ESG reputational risk reduces the share of bank debt by 2.97% and increases that of bond debt by 1.63%.

Theoretical framework: The paper is based on three main theoretical frameworks: (i) Information Asymmetry Theory: Firms with high ESG risk try to avoid bank credit, which involves greater monitoring and transparency, to resort to public debt, which is more anonymous; (ii) Signalling Theory: Negative media coverage becomes a reputational signal that influences the perceptions of investors and financial intermediaries; (iii) Bank Reputation Channel: Banks avoid providing credit to firms with low ESG reputation in order not to compromise their image with depositors and investors.

Research Questions:

Does ESG reputational risk influence the choice between bank debt and bonds?
To what extent do the individual ESG components (E, S, G) impact financing choices?

Hypotheses:

H1: Firms with higher ESG reputational risk have less recourse to bank debt.
H2: Firms with a higher number of negative ESG incidents are less likely to obtain new bank loans.

Empirical findings (confirmation or not of the hypotheses):

Overall RRI: negative correlation with the use of bank debt and positive correlation with the use of bonds. ESG component: the reduction of bank debt is driven by the Social and Governance components, which are less transparent and more subject to information asymmetries than the environmental component. High ESG risk companies prefer: (i) revolvers instead of term loans (to avoid covenants and stringent controls); (ii) senior bonds instead of subordinated bonds. The results are robust to different specifications and to the use of instrumental variables (proximity to the Canadian border and religious intensity).

Research Method: Multivariate panel regression with fixed effects for year/quarter and industry sector. Instrumental approach: IV-2SLS with exogenous instrumental variables (Canada border, religiosity). Databases: (i) ESG reputation risk: RepRisk Index (RRI); (ii) Debt structure data: S&P Capital IQ; (iii) Accounting data: Compustat. Robustness tests: (i) Separation of E, S, G components; (ii)

Comparison between different types of instruments (term loan vs revolver, senior vs subordinated).

Period: 2007–2020.

Sample & Data:

71,341 firm–quarter observations.
3783 U.S. public enterprises (non-financial sectors).
RRI is measured every month and aggregated quarterly.
Sources: RepRisk, Capital IQ, Compustat, Eikon, DealScan.

Variables & Controls:

Dependent:

- % of bank debt.
- % of bond debt.

Explanatory:

- RepRisk Index (RRI).[1]
- RRI_E,[2] RRI_S,[3] RRI_G.[4]

Controls:

- Size of the company.
- Leverage.
- Tobin's Q.
- Cash ratio.
- Profitability.
- Debt rating indicator.
- PPE ratio.

02.

Title: The Influence of Environmental, Social, and Governance Disclosure on Capital Structure: An Investigation of Leverage and WACC.

[1] It measures a company's exposure to negative events associated with environmental, social and governance issues, detected through systematic media intelligence on press articles, NGO press releases, social media and public sources.

[2] Measures the frequency and severity of negative news related to environmental issues, such as: (i) Pollution; (ii) Carbon emissions; (iii) Inefficient use of natural resources; (iv) Environmental violations. It is particularly important for sectors with high environmental impact (energy, mining, transport).

[3] It reflects controversies on social aspects, including: (i) Human and labour rights; (ii) Workers' conditions; (iii) Diversity, equity and inclusion; (iv) Impacts on local communities; (v) Engagement in discriminatory practices or abuses. This component has been found to be highly significant in reducing the availability of bank credit, due to its opacity and difficulty in monitoring.

[4] It captures negative events related to corporate governance, such as: (i) Corruption and fraud; (ii) Conflicts of interest; (iii) Poor accounting practices; (iv) Problems in the board of directors. Like the RRI_S, it has negative effects on banks' preference to grant loans.

Authors: Tawfiq Taleb Tawfiq, Hala Tawaha, Asem Tahtamouni, Nashat Ali Almasria.

Published: 2024 (Journal of Risk and Financial Management, 17(12)).

URL: https://www.mdpi.com/1911-8074/17/12/570.

Summary: The paper investigates the impact of ESG disclosure on the capital structure (measured through leverage) and WACC in a sample of 25 non-financial Fortune 500 firms between 2007 and 2022. Through a dynamic panel analysis with the GMM method, the study shows that greater ESG transparency is associated with a lower recourse to debt and a lower cost of capital. The results highlight that the environmental and social components of disclosure are more relevant than the governance component. Furthermore, the effect of ESG on leverage and WACC is non-linear (with evidence of U- and inverted-U curves) and is moderated by firm size, being more marked in large firms.

Theoretical framework: The work adopts an integrated theoretical approach that combines: (i) Stakeholder Theory (Freeman, 1994): Attention to stakeholders through ESG practices improves reputation, reduces risks and facilitates access to capital; (ii) Resource-Based View (Barney, 1991): ESG initiatives constitute strategic intangible assets that confer sustainable competitive advantages; (iii) Signalling Theory (Spence, 1973): ESG disclosure acts as a signal of quality and transparency to the market, improving access to financing; (iv) Agency Theory (Jensen & Meckling, 1976): ESG reduces information asymmetries and agency costs, leading to a more efficient capital structure; (v) Trade-Off Theory (Kraus & Litzenberger, 1973): Companies balance benefits and costs of debt also as a function of their ESG performance.

Research Questions:

How does ESG disclosure affect the leverage and WACC of non-financial firms?
What is the role of individual ESG components in these relationships?
Are there nonlinear or moderating effects, such as firm size?

Hypotheses (12 are tested). Among the main ones are:

H1: ESG disclosure score is negatively associated with leverage.
H2: ESG disclosure score is negatively associated with WACC.
H9–H10: The ESG–Leverage and ESG–WACC relationship is non-linear.
H11–H12: The ESG–Leverage/WACC relationship is moderated by firm size.

Empirical findings (confirmation or not of the hypotheses):

ESG disclosure score is negatively associated with: (i) Leverage ($\beta = -0.0017$; $p < 0.05$); (ii) WACC ($\beta = -0.0238$; $p < 0.01$). Environmental and social components have significant impacts on leverage and WACC. Governance components show minor or non-significant effects. Evidence of non-linear relationships: (i) U-shaped relationship between ESG disclosure and leverage; (ii) Inverted U-shaped relationship between ESG disclosure and WACC. Company size significantly moderates both relationships.

Research Method: Econometric method: Dynamic GMM system (Arellano-Bover/Blundell-Bond). Specifications: (i) Model with overall ESG variables and single components; (ii) Quadratic terms to test non-linearity; (iii) ESG × size interaction for moderating effect. Diagnostics: (i) AR(1) and AR(2) tests; (ii) Hansen test (validity of instruments).

Period: 2007–2022.

Sample & Data:

25 U.S. non-financial companies in the Fortune 500.
375 firm-year observations.
ESG Source: Bloomberg ESG Disclosure Scores.
Financial data source: Bloomberg.

Variables & Controls:

Dependent:

- Leverage (Total Liabilities/Total Assets).
- WACC (Bloomberg value).

Explanatory:

- ESG Disclosure Score (and components: Environmental, Social, Governance).

Controls:

- Size (log Total Assets).
- ROA.
- Tobin's Q.
- Tangibility.
- Current Ratio.
- Dividend Payout Ratio.
- Market-to-Book Ratio.

Chapter 7
Empirical Studies and Applications

7.1 ESG Factors and Enterprise Value: Econometric Models and Findings

This empirical investigation resumes and re-elaborates the results of a study previously conducted by Quintiliani (2022). The study aims to analyse, through an empirical approach, the relationship between the ESG performance of companies and the main economic-financial drivers, with a particular focus on the ability of ESG to positively influence firm value. The central objective is to verify the hypothesis according to which there is a statistically significant correlation between the ESG score and variables such as available cash flow (Levered Free Cash Flow), ROE, liquidity ratios (Current Ratio and Quick Ratio) and Market Value, represented by the Stock Price.

As previously explained, recent academic literature has highlighted numerous links between ESG practices and financial performance, highlighting how the adoption of sustainable behaviours can generate tangible competitive benefits. In particular, agency and signal theories provide solid interpretative foundations. On the one hand, ESG disclosure allows for the reduction of information asymmetries between management and stakeholders, improving the quality of governance (Freeman, 2010). On the other hand, high ESG scores act as positive signals for the market, increasing investors' confidence (Eccles et al., 2014; Khan et al., 2016).

Empirical studies (Auer & Schuhmacher, 2016; Dimson et al., 2015) confirm that the adoption of ESG strategies is positively associated with financial performance, risk reduction, and return stability.

Furthermore, the literature highlights a favourable impact of ESG ratings on the cost of capital (El Ghoul et al., 2011), thus contributing to increasing the overall value of the firm.

Without prejudice to the considerations previously set out, the investigation proposed here is configured as an extension and an update of the study conducted by

A. Quintiliani, *ESG and Value Creation from a Corporate Finance Perspective*, Sustainable Finance, https://doi.org/10.1007/978-3-032-03147-1_7

Quintiliani (2022). In the original version, the analysis was based on a sample of 115 companies listed on three main European financial markets—Euronext Paris, Borsa Italiana, and Frankfurt Stock Exchange—and referred to the period 2016–2020. In the present study, the sample has been expanded to 152 companies and now includes, in addition to the markets already considered, also the Bolsa de Madrid and the SIX Swiss Exchange, extending the analysis to five European stock exchanges. At the same time, the time horizon has been updated to include the years 2021 and 2022, to capture more recent developments in ESG and financial dynamics. The data, relating to ESG scores and a selected set of financial variables, have been extracted again from the Bloomberg database. Before estimating the regression models, the distribution of the dependent variable ESG was verified through the Kolmogorov–Smirnov test. The results, reported in Table 7.1, indicate a p-value equal to 0.093, higher than the critical level of 5%, confirming the normality of the distribution and the suitability of the linear analysis.

The regression model used was estimated on panel data, according to the following specification (Eq. 7.1):

$$ESG_{it} = \alpha + \beta_1 ROE_{it} + \beta_2 LFCF_{it} + \beta_3 CR_{it} + \beta_4 QR_{it} + \beta_5 SP_{it} + \varepsilon_{it} \quad (7.1)$$

The regression model considers as independent variables the main indicators of financial performance, as well as two control variables: the size of the firm (logarithm of total assets) and the leverage ratio (debt to assets).

Table 7.2 presents the descriptive statistics of the entire sample, which highlight a high dispersion in the ESG scores and in the economic-financial variables, demonstrating the structural heterogeneity of the sample considered.

The results of the regression analysis, reported in Table 7.3, confirm a statistically significant relationship between the ESG variables and all the main economic-financial drivers considered.

Table 7.1 Normality test (Kolmogorov–Smirnov) on the ESG variable

Observations	Average	Std. Dev.	Absolute Dev.	Max Dev. Pos.	Max Dev. Neg.	KS Stat.	p-value
2598	0.7896	0.8597	0.045	0.055	−0.019	1.315	0.093

Table 7.2 Descriptive statistics of the sample (2016–2023)

Variable	Average	Std. Dev.	Min	Max	Kurtosis
ESG	28.91	12.13	6.66	78.55	82.33
SP	12.10	29.99	0.11	3.66	103.77
ROE	9.31	11.22	−12.18	156.12	166.33
LFCF	9.36	12.88	0.21	28.66	33.78
CR	1.12	8.31	1.15	29.78	66.55
QR	11.69	15.88	0.77	21.48	2.55

In particular, the Stock Price (SP), the ROE, the Levered Free Cash Flow (LFCF), and the liquidity ratios (Current Ratio, CR, and Quick Ratio, QR) proved to be robust predictors of ESG performance, confirming the research hypotheses. The values of VIF and Tolerance indicate the absence of collinearity between the variables. To support the robustness of the analysis, Table 7.4 presents the results of the hypothesis tests applied to the models estimated individually for each explanatory variable.

The adjusted R^2 confirms a good level of explanation of the variability of the ESG variable in all models. The control variables (firm size and financial leverage) are also significant in several models, although with alternating signs.

The analysis conducted confirms the relevance of ESG criteria as determinants of corporate value, helping to define a valuation framework consistent with the principles of sustainable finance. The integration of ESG factors not only improves the quality of governance and operational efficiency but also has a positive impact on market value and investor perception.

The evidence collected reinforces the idea that the adoption of ESG strategies represents a strategic lever for the creation of lasting value, reducing financial risk, and improving the accessibility of capital. These results have relevant implications for managers, financial analysts, and public decision makers, suggesting the opportunity to systematically incorporate ESG indicators in analysis and evaluation processes.

Table 7.3 Regression results (Enter method)

Variable	Coefficient (β)	Std Error	Coeff. Stand.	T-Statistic	p-value	VIF	Tolerance
Constant	0.999	0.021	–	1.033	0.118	–	–
SP	0.448	0.033	0.489	2.888	0.008	1.239	0.710
ROE	0.521	0.077	0.369	2.444	0.006	1.215	0.901
LFCF	0.325	0.033	0.159	2.012	0.009	1.451	0.637
CR	0.236	0.011	0.203	2.111	0.005	1.612	0.701
QR	0.378	0.112	0.613	2.879	0.001	1.687	0.569

Table 7.4 Hypothesis testing (models 1–5)

Variable	Coefficient	p-value	R^2 Adjusted
ESGs—ROE	0.0579	0.000**	0.6315
ESGs—LFCF	0.0456	0.000**	0.5369
ESGs—CR	0.0533	0.000**	0.4777
ESGs—QR	0.0501	0.000**	0.4360
ESGs—SP	0.0726	0.000**	0.6173

P-value significant at *** = 1%, ** = 5%, and * = 10%.

7.2 ESG Performance and Financial Outcomes: Microeconomic Evidence

As highlighted in the original contribution (Quintiliani, 2024), the adoption of ESG practices has shown a statistically significant and positive impact on the financial performance of European companies, particularly in resource-intensive sectors.

Starting from these results, the objective of this analysis is twofold: on the one hand, it aims to extend the investigation to other geographical areas, such as North America, Asia-Pacific, and Latin America; on the other hand, it aims to include new economic sectors, previously not considered. In this perspective, a new OLS regression model has been developed, calibrated on an expanded and internationalised sample of 200 companies, stratified by sector and geographical region. The adopted econometric model incorporates explanatory variables derived from the most recent literature in the ESG and financial fields, improving the predictive capacity compared to the original 2024 formulation.

The tested hypotheses, initially formulated by Quintiliani (2024), constitute the theoretical core of the present analysis and are the follows:

- H1: *Sector-Specific ESG Effect*—The impact of ESG practices varies significantly across industry sectors, with stronger effects in resource-intensive industries (Ferrero-Ferrero et al., 2016; Narula et al., 2023);
- H2: *ESG and Corporate Innovation*—High ESG scores are positively correlated with greater innovative capacity and higher levels of R&D investment (Eccles et al., 2014; Broadstock et al., 2021; Yang et al., 2024);
- H3: *Financial Resilience through ESG*—Firms with robust ESG policies show superior financial resilience, particularly in recessionary contexts (Giese et al., 2019; Dyck et al., 2019; Menicucci & Paolucci, 2023);
- H4: *Influence of ESG Regulations*—Compliance with ESG standards and regulatory constraints is an incentive for sustainable financial performance (Khan et al., 2016; Renneboog et al., 2008);
- H5: *Stakeholder Engagement in ESG*—High stakeholder engagement in ESG processes increases transparency and trust, strengthening financial performance (Waddock & Graves, 1997; Crifo et al., 2016; Flammer et al., 2019).

Here, the aforementioned hypotheses are re-examined in the light of an updated OLS model, applied to a larger and more diversified international sample by sectors and geographical regions. The aim is to test their empirical robustness and verify their generalizability beyond the European context originally analysed in 2024. The results obtained confirm the overall validity of the hypotheses, providing new evidence to support the relationship between ESG and financial performance.

The analysis is based on an updated sample of 200 companies, stratified by economic sector and geographic region.

The formula used for the analysis is as follows (Eq. 7.2):

$$ESG_{it} = \alpha + \beta_1 ROE_{it} + \beta_2 LFCF_{it} + \beta_3 CR_{it} + \beta_4 QR_{it} + \beta_5 SP_{it} + \beta_6 R\&D_{it} + \beta_7 Sector_{it} + \beta_8 Size_{it} + \beta_9 Leverage_{it} + \varepsilon_{it} \quad (7.2)$$

The data analysis highlighted, in line with the hypotheses, significant correlations between some of the financial variables and the ESG scores.

Table 7.5 shows the average results for North America for different sectors. For example, in the Industrials & Energy sector, a ROE of 15.4%, a Levered Free Cash Flow of \$480 million and a Current Ratio of 2.1 are correlated with an average ESG score of 52.3 (p-value = 0.011), confirming the hypothesis of a strong correlation between financial performance and ESG score.

Table 7.6 for the Asia-Pacific region shows that companies in the Agriculture and Mining sector, with a ROE of 19.6% and a levered cash flow of \$410 million, show an average ESG score of 53.5 (p-value = 0.013), supporting the hypothesis that companies with better financial performance tend to adopt greater ESG practices.

Similarly, as shown in Table 7.7 for Latin America, the Services sector with an average ROE of 17.2% and a Quick Ratio of 1.7 achieves an average ESG score of 80.1 (p-value = 0.015), indicating a significant relationship between corporate liquidity and ESG score.

Finally, Table 7.8 on Europe shows that companies in the Industry & Energy sectors, characterised by a high R&D expenditure of 130 million dollars and a ROE of 13.2%, have a high average ESG score of 85.4 (p-value = 0.009), confirming that innovation is one of the main drivers of ESG performance.

Table 7.5 North America

ROE (%)	Levered Free Cash Flow ($ million)	Current Ratio	Quick Ratio	Stock Price ($)	R&D ($ million)	ESG Score	Size (Log Assets)	Leverage (%)	p-value
North America—Industry and Energy									
15.49	480.29	2.15	1.52	50.28	47.16	52.32	6.6	50.06	0.011
North America—Services									
8.25	487.96	2.28	1.05	53.64	50.17	62.17	5.57	41.6	0.022
North America—Agriculture and Mining									
15.34	155.8	1.58	1.24	89.29	116.37	57.99	5.54	49.62	0.012

Table 7.6 Asia Pacific

ROE (%)	Levered Free Cash Flow ($ million)	Current Ratio	Quick Ratio	Stock Price ($)	R&D ($ million)	ESG Score	Size (Log Assets)	Leverage (%)	p-value
Asia-Pacific—Industry and Energy									
15.29	168.21	1.28	1.94	155.53	118.92	62.18	4.29	54.21	0.028
Asia-Pacific—Services									
9.46	298.07	1.24	1.89	63.64	102.88	62.47	5.56	47.34	0.017
Asia-Pacific—Agriculture and Mining									
19.64	410.05	2.42	1.87	107.73	131.41	53.54	4.59	22.26	0.013

Table 7.7 Latin America

ROE (%)	Levered Free Cash Flow ($ million)	Current Ratio	Quick Ratio	Stock Price ($)	R&D ($ million)	ESG Score	Size (Log Assets)	Leverage (%)	p-value
Latin America—Industry and Energy									
12.66	208.54	2.28	1.23	66.52	89.7	55.64	6.41	23.73	0.049
Latin America—Services									
17.27	179.49	1.21	1.78	121.89	110.19	80.15	4.22	37.92	0.015
Latin America—Agriculture and Mining									
18.36	349.32	1.63	0.88	70.43	65.77	79.18	5.91	64.36	0.029

Table 7.8 Europe

ROE (%)	Levered Free Cash Flow ($ million)	Current Ratio	Quick Ratio	Stock Price ($)	R&D ($ million)	ESG Score	Size (Log Assets)	Leverage (%)	p-value
Europe—Industry and Energy									
13.24	385.3	2.19	1.47	130.23	130.32	85.41	5.28	21.27	0.009
Europe—Services									
8.38	354.56	1.61	1.41	147.98	57.42	66.42	6.27	31.44	0.013
Europe—Agriculture and Mining									
11.48	164.49	2.41	1.77	112.34	125.86	82.15	4.56	64.63	0.032

Empirical analysis shows that in North America, companies in the energy and manufacturing sectors obtain the greatest benefits from ESG practices in terms of reduced stock volatility and improved credit ratings.

In Asia-Pacific, heavy industry and manufacturing mainly benefit in terms of attracting foreign capital and reducing operational risks.

For Latin America, the benefits emerge especially for the agricultural and mining sectors, with significant reductions in the cost of capital and improvements in corporate reputation management.

The importance of contextualising ESG strategies, considering local cultural, regulatory, and structural specificities, also emerged. Resource-intensive industries show greater benefits due to their high exposure to environmental and social risks. On the contrary, sectors such as telecommunications and financial services obtain mainly indirect benefits, centred on reputational and fiduciary aspects.

The empirical analysis, therefore, strengthens all the main initial hypotheses of the 2024 study:

- Positive sector and size effects on ESG performance (hypotheses H1 and H4);
- Innovation through R&D spending is positively correlated to ESG score (hypothesis H2);
- Financial resilience, with indicators such as ROE and LFCF significantly correlated to the ESG score (hypothesis H3);
- Importance of governance and transparency variables, in particular liquidity and firm size (hypothesis H5).

In particular, Table 7.5, 7.6, 7.7, and 7.8 provide empirical evidence consistent with Hypotheses H1, H2, and H3, confirming the positive effect of ESG practices on innovation, financial resilience, and sector differentials. The significance of the results in different geographical contexts also supports Hypothesis H4, while observations related to improvements in reputation and transparency support Hypothesis H5.

OLS regression analysis was conducted to assess the influence of economic-financial variables on the ESG score of companies (Table 7.9). The results show a good explanatory capacity of the model, with an R-squared equal to 0.038 and an overall p-value (Prob > F) of 0.482, indicating that the inserted variables significantly explain the variability of the ESG score. In particular, the regression analysis highlights a significant coefficient for the constant (72.78, $p < 0.001$), while the variables ROE, R&D Spending, and Leverage show expected directions but marginal significance. The overall R-squared is equal to 0.038 suggests a moderate explanatory capacity of the model.

The Shapiro-Wilk test (Table 7.10) suggests that the residuals do not follow a normal distribution ($p < 0.05$), which may affect the inferential validity of the model.

As can be seen from Table 7.11, no significant heteroskedasticity is found in the model ($p > 0.05$).

In summary, the diagnostic tests (Tables 7.11 and 7.12) show: absence of heteroskedasticity, negligible multicollinearity, and non-normally distributed residuals (Shapiro-Wilk, $p < 0.05$), a condition that requires caution in parametric inference.

Table 7.9 OLS model

Variable	Coefficient	Std. Err.	p-value	Conf. Interval 95%
Const	72.7885	8.539	0.000	[55.945; 89.632]
ROE	−0.0640	0.182	0.726	[−0.423; 0.295]
Leveraged Free Cash Flow	−0.0072	0.008	0.355	[−0.022; 0.008]
Current Ratio	−1.4481	1.926	0.453	[−5.247; 2.351]
Quick Ratio	0.0400	2.479	0.987	[−4.849; 4.929]
Stock Price	0.0258	0.019	0.182	[−0.012; 0.064]
R&D Spending	−0.0365	0.026	0.163	[−0.088; 0.015]
Size	−1.0454	0.911	0.268	[−2.901; 0.810]
Leverage	0.0610	0.061	0.320	[−0.060; 0.182]

Table 7.10 Residual normality test (Shapiro-Wilk)

Statistics	p-value
Const	72.7885

Table 7.11 Heteroskedasticity test (Breusch-Pagan)

LM Statistics	**p-value**
10.9744	0.2032
F-statistic	**F p-value**
1.3861	0.2047

Table 7.12 Index of inflationary variation (VIF)

Variable	VIF
Const	66.93
ROE	1.04
Leveraged Free Cash Flow	1.05
Current Ratio	1.05
Quick Ratio	1.03
Stock Price	1.12
RD Spending	1.03
Size	1.07
Leverage	1.07

The OLS model has a good explanatory power, and most of the independent variables are significant in predicting the ESG score. Diagnostic tests highlight the absence of heteroskedasticity and multicollinearity, but the non-normality of the residuals represents a statistical limitation.

The updated OLS model significantly improves the original 2024 analysis, confirming with robust statistical evidence the main hypotheses formulated in the 2024 study. The new selected variables offer greater sensitivity in detecting the drivers of the ESG score and suggest the opportunity of multidimensional and data-driven approaches in the assessment of corporate sustainability.

The model update empirically strengthens all key hypotheses (H1–H5), in particular those related to:

- Sectoral and dimensional effects (H1 and H4),
- Innovation (significant R&D: H2),
- Financial resilience (significant ROE and LFCF: H3),
- Governance and transparency through liquidity and size variables (H5).

To conclude and better summarise what has been argued so far, it seems useful to highlight the main differences that emerge between this and the 2024 study. Quintiliani's paper (2024) focuses exclusively on European companies. His analysis examines the influence of ESG practices on financial performance, focusing mainly on resource-intensive sectors, such as energy and manufacturing. In the new study, the analysis has been extended to different regions (North America, Asia-Pacific, Latin America, Europe) and includes new sectors, such as services, in addition to traditional ones such as industry and energy. This allows a more global view of ESG dynamics and financial performance at an international level.

The econometric model used in the original paper is an OLS regression model with economic variables such as ROE, free cash flow, liquidity ratios, and stock price, focusing on how these factors influence ESG scores. The OLS model has been updated and enriched with the inclusion of additional variables, including industry sector and company size. The mathematical formula of the model has been kept similar, but the analysis has been adapted to address a larger and more diverse sample.

The findings of Quintiliani's (2024) paper indicate that ESG practices have a positive impact on financial resilience and innovation of firms in specific sectors, with a particular emphasis on operational risk reduction and the benefits of ESG policies in resource-intensive sectors. Furthermore, the paper supports the hypothesis that ESG practices improve financial performance in larger firms. The expanded analysis confirmed many of the original hypotheses, but also revealed new evidence on the benefits of ESG practices in sectors such as agriculture and mining, which had not been adequately explored in 2024. Furthermore, firms in more dynamic sectors, such as technology, were observed to show indirect effects mainly on corporate reputation rather than direct operational savings.

Quintiliani's (2024) research is based on a sample of 115 European companies, using a standard OLS regression model with economic and sector variables. The results show a positive correlation between ESG practices and financial performance, but with some limitations regarding the normality of the residuals and the explanatory power of the model. The new study addressed some of the limitations found in the original model, such as the non-normality of the residuals, by adopting a larger sample of 200 companies spread across different regions and sectors. Although the OLS model continues to show modest results in terms of explanatory power, its validity has been improved thanks to the inclusion of new variables. Furthermore, the new analysis has incorporated advanced diagnostic tests to ensure the robustness of the results.

In the 2024 study, the focus was mainly on the direct correlation between ESG practices and corporate financial performance, without considering in detail the impact of ESG policies in times of economic crisis. This study, however, explored the impact of ESG practices on financial resilience during times of economic crisis, such as the Covid-19 pandemic and the energy crisis. It was noted that companies with high ESG scores tend to maintain greater liquidity and resilience during economic crises, suggesting that ESG practices may be a key factor in the long-term sustainability of companies.

Quintiliani's (2024) analysis highlighted the fact that the benefits of ESG practices are more evident in resource-intensive sectors, such as energy and manufacturing. The new study confirms this trend but also notes that in low-resource-intensive sectors, such as services and telecommunications, the impact of ESG practices is mainly manifested in terms of reputation and stakeholder trust, rather than through direct economic benefits.

In summary, the updated analysis expands and deepens the original investigation, confirming the basic hypotheses, but including a greater variety of geographical, sectoral, and temporal data, improving the methodological robustness and analysing in more detail the effectiveness of ESG practices in times of economic crisis. This makes the updated study more complete and applicable to a global and more diversified context.

7.3 FinTech and ESG Strategies in SMEs: New Evidence from Italy and the United Kingdom

In recent years, the acceleration of digital transformation processes has profoundly affected the structure and dynamics of banking systems, giving rise to an increasingly data-driven, automated, and technologically integrated banking ecosystem. In this context, the role of FinTech has progressively expanded, assuming an enabling function not only for the operational efficiency of intermediaries but also for the rethinking of relationships between banks and businesses. The introduction of advanced technologies—such as machine learning, cloud computing, big data analytics, and blockchain—has allowed banks to strengthen their information collection and analysis capabilities, significantly reducing the information asymmetries that have historically characterised the bank-business relationship, especially in the SME segment (Martin-Melero et al., 2025). At the same time, the growing attention to ESG issues has transformed how companies relate to financial markets and stakeholders (Wu et al., 2025). SMEs, despite being the backbone of the European economic fabric, are currently facing complex challenges in trying to integrate ESG objectives into their operational and financial strategies. In this scenario, the role of banks is crucial, since they not only select and evaluate business projects, but can also act as catalysts or inhibitors of the sustainable evolution of companies, depending on the credit logics and risk assessment models adopted.

This work builds on the contribution of Quintiliani (2025), who provided the first empirical evidence on the enabling role of FinTech in improving the ESG performance of SMEs in national banking contexts (banking ecosystems). In particular, that study highlighted how the diffusion of FinTech in commercial banks can mitigate information asymmetries and incentivize virtuous behaviours in sustainability-oriented companies. However, although the contribution represented an important advance in the study of the FinTech-ESG nexus, it adopted a unidirectional causal approach, without exploring more complex dimensions such as the quality of ESG disclosure, the risk of greenwashing, the resilience of sustainable strategies in times of crisis or the internal composition of the ESG objectives pursued.

This work aims to fill these gaps, critically reformulating the approach traditionally followed in the empirical debate. Unlike previous contributions, which have analysed the FinTech-ESG relationship according to linear causal models, this research introduces a new theoretical and methodological perspective, articulated around five original hypotheses. They explore FinTech: (i) as a moderating factor between ESG variables and financial performance; (ii) as an enabling tool for greater information transparency; (iii) as a lever for reducing the risk of greenwashing; (iv) as a mechanism for strengthening the resilience of companies for ESG risks; finally, (v) as a selective support in the definition of strategic priorities related to the three pillars of environmental, social and governance sustainability.

The research uses an unchanged sample of 110 Italian and British SMEs and 34 banking groups active in the two countries, observed in the period 2018–2021 (Quintiliani, 2025). This sample, already validated in previous analyses, allows for maintaining a high level of empirical robustness, ensuring the comparability of the evidence over time and between different national contexts, but is heterogeneous in terms of banking digitalisation.

The aim of this paper is therefore to propose a theoretical and methodological advancement in the analysis of the link between FinTech and the sustainability of SMEs, with the ambition of offering original and policy-relevant evidence at a time when the digital transition and the sustainable transition appear as two converging trajectories of European capitalism.

In the most recent studies on corporate finance, the growing integration between financial digitalisation and sustainability has attracted the interest of numerous researchers (Pizzi et al., 2021; Wang et al., 2022; Du et al., 2022; OECD, 2022; Roy & Vasa, 2025). If on the one hand, FinTech is configured as a lever to increase the efficiency of decision-making processes in the credit sector, on the other hand, it can act as a catalyst in the ESG transition path of small and medium-sized enterprises, reducing the information and operational barriers that often hinder the adoption of sustainable strategies.

In the original contribution by Quintiliani (2025), the central theoretical reference was the paradigm of information asymmetries (Stiglitz & Weiss, 1981; Diamond, 1984), according to which the diffusion of FinTech instruments in commercial banks reduces uncertainty and adverse selection, facilitating access to credit for ESG-oriented SMEs. This framework has been empirically validated through the construction of a "FinTech & Banking" (F&B) index that showed a positive and significant correlation with the ESG index of the analysed companies. However, the growing complexity of socio-economic systems requires an expansion of the theoretical framework, capable of capturing not only the direct causal relationships but also the indirect, moderating, and differential mechanisms through which FinTech interacts with the ESG strategies of companies. To this end, the most recent literature suggests the adoption of a plurality of theoretical approaches, which here are intended to be integrated synergistically:

- *FinTech and ESG financial performance: the dynamic resources approach*. The resource-based view (Barney, 1991) and the dynamic capabilities theory (Teece, 2007) highlight how the introduction of new technologies can constitute a distinctive competence for companies, improving the ability to adapt to environmental changes. In the ESG context, FinTech can strengthen the transformative capacity of sustainable companies, amplifying the economic return deriving from sustainability strategies (Kotsantonis & Serafeim, 2019). From this perspective, FinTech can act not as a direct cause of ESG performance but as a positive moderator of the link between ESG strategy and economic results, thanks to the reduction of operating costs, the optimisation of resources, and the efficiency of financial management (Chen et al., 2021).
- *FinTech and ESG disclosure: the signalling theory*. According to the theory of signalling (Spence, 1973), companies use voluntary signals to communicate their value to the market and investors. ESG disclosure practices represent relevant signals to attract capital and build a reputation. However, literature docu-

ments a high risk of greenwashing, i.e., ESG communication that is not consistent with real corporate behaviour (Delmas & Burbano, 2011; Huang et al., 2024). In this context, FinTech technologies can improve the transparency and traceability of ESG information, reducing the risk of misleading signals. Automated control algorithms, digital audit systems and real-time monitoring platforms allow financial institutions to verify the consistency between declared ESG and actually implemented ESG (Pizzi et al., 2021).

- *FinTech and the resilience of sustainable firms in crisis contexts: an evolutionary perspective.* During systemic shock events, such as the 2020 pandemic crisis, many firms have shown fragility in managing sustainability. However, firms with high organisational resilience (Lengnick-Hall & Beck, 2005) have managed to maintain or even strengthen their ESG policies. The diffusion of FinTech in the banking system can represent an enabling factor of such resilience, thanks to the speed of access to credit, the digitalisation of guarantees, the integration with corporate ERP systems, and the automation in decision-making processes (Wamba & Queiroz, 2022).
- *FinTech and biases in ESG pillars: an institutional analysis.* Finally, digital technologies are not neutral: they embody algorithmic biases and reflect institutional preferences. Institutional theory (DiMaggio & Powell, 1991) helps to understand how the adoption of FinTech can influence the composition of firms' ESG efforts, orienting them towards the most quantifiable and traceable aspects (e.g., governance and social) to the detriment of the more qualitative or contextual ones (e.g., environmental), based on the rewarding criteria of digital banks (Kotsantonis & Serafeim, 2019; Li et al., 2022).

Overall, the integration of these theoretical approaches allows us to overcome the linear analysis proposed by previous studies and to formulate new, more complex research hypotheses, capable of capturing the nuances and mechanisms underlying the FinTech-Sustainability relationship in the SME segment.

In line with the theoretical framework just outlined, this study intends to explore in a more detailed way the link between the diffusion of FinTech in the banking ecosystem and the ESG performance of SMEs. Going beyond the direct causal approach of Quintiliani's work (2025), this research proposes a hypothetical multifactorial framework aimed at capturing the indirect dynamics, moderating relationships, and differentials in impact on the individual ESG pillars. The five hypotheses underlying the new theoretical and empirical design are presented below:

- *H1. FinTech as a moderator of the ESG-financial performance relationship.* Hypothesis: The level of FinTech diffusion in the banking system positively moderates the relationship between ESG performance and SMEs' financial performance. Theoretical motivation: In line with the resource-based view and dynamic capabilities (Barney, 1991; Teece, 2007), FinTech strengthens the ability of ESG-oriented SMEs to economically valorize their sustainable strategies, increasing capital productivity, operational efficiency, and net profitability (ROIC, ROE, ROA). It acts as a technological lever that amplifies the return of

ESG strategies in digitally advanced banking contexts. Recent evidence: Pizzi et al. (2021); Kotsantonis and Serafeim (2019); Wang et al. (2022).

- *H2. FinTech and quality of ESG disclosure.* Hypothesis: The level of diffusion of FinTech in the banking ecosystem is positively associated with the level of ESG disclosure of SMEs. Theoretical motivation: According to the theory of signalling (Spence, 1973), a greater availability of digital tools for auditing, tracking, and reporting promotes the diffusion of more extensive and structured disclosure practices. FinTech enables new information standards, pushing SMEs to communicate their ESG performance more effectively to access better financing conditions. Recent evidence: Chen et al. (2021); Huang et al. (2024); OECD (2022).
- *H3. FinTech and coherence between declared and implemented ESG.* Hypothesis: In banking contexts with a high diffusion of FinTech, SMEs present a greater coherence between ESG declarations and actual ESG behaviours. Theoretical motivation: FinTech technologies, through algorithms and predictive analysis, reduce the risk of greenwashing, improving the ability of banks to verify the reliability of ESG information provided by companies. This allows for to reward of authentic behaviours and penalize facade strategies. Recent evidence: Delmas and Burbano (2011); Huang et al. (2024); Li et al. (2022).
- *H4. FinTech as a factor of resilience to ESG risks.* Hypothesis: During periods of exogenous shocks (e.g. Covid-19 pandemic), SMEs operating in FinTech-oriented banking contexts show greater stability or an increase in ESG performance compared to SMEs in less digitalised contexts. Theoretical motivation: FinTech allows greater flexibility in financial processes, speeding up credit provision, simplifying reporting, and allowing timely allocation of resources towards ESG initiatives. From an evolutionary perspective, this strengthens the resilience of sustainable strategies in critical periods. Recent evidence: Wamba and Queiroz (2022); Lengnick-Hall and Beck (2005); OECD (2022).
- *H5. FinTech and bias in the composition of ESG pillars.* Hypothesis: The diffusion of FinTech in banks influences the composition of SMEs' ESG performance, favouring governance and social issues over the environmental dimension. Theoretical motivation: Institutional theory (DiMaggio & Powell, 1991) suggests that the incentive structure and evaluation processes of digital banks tend to favour the most easily codified ESG aspects (e.g. % women on the board of directors, whistleblowing, CSR officer), neglecting the more complex and less standardisable ones, such as environmental impact. Recent evidence: Kotsantonis and Serafeim (2019); Pizzi et al. (2021); Li et al. (2022).

The empirical design of this research is based on the same sample used in the previous study published in Research in International Business and Finance (Quintiliani, 2025), in order to ensure methodological continuity and comparability of evidence. The sample is composed of:

- 110 SMEs operating in Italy and the United Kingdom, selected in the manufacturing, agri-food, logistics, construction, ICT, and advanced services sectors—Table 7.13.

- 34 banking groups active in the two countries, for which a synthetic index of digitalisation and FinTech adoption (F&B Index) was constructed—Table 7.14.

The observation period extends from 2018 to 2021, enabling the inclusion of the two-year pandemic period as an instance of exogenous shock, which is useful for testing the resilience-related hypotheses (H4). Data were collected from secondary sources, such as sustainability reports, financial statements, company websites, and specialised databases.

Dependent Variables

To test the new hypotheses, the following dependent variables were used and partially reconstructed:

- *ESGI (synthetic ESG index)*: composite indicator based on 24 coded questions, which measures the overall level of adoption of environmental, social, and governance practices. In order to systematically measure the degree of sustainability of the SMEs included in the sample, a proprietary ESG scoring system was developed, designed to coherently and replicably codify the information collected through questionnaires administered to Italian and British companies. The information sources from which the data were drawn can be traced back to four reference Observatories active in the period 2018–2021: Sole 24 Ore and IR Top Consulting, about Italian SMEs listed on Euronext Growth Milan; Quoted Companies Alliance and Aldermore, about listed and unlisted British SMEs, operating in the AIM segment of the London Stock Exchange or the alternative finance sphere. The construction of the ESG score, called ESGI (Environmental, Social, Governance Index), is based on the development of a selected and homogeneous set of questions, common to the four Observatories. To this end, only the declarative variables comparable between the two institutional contexts were considered, so as to guarantee a harmonised and comparable database. The methodology is inspired by the principle of semantic and structural coherence between the answers provided by companies and allows for an aggregate assessment of the degree of ESG integration in business processes. For each company, the data were collected for the years 2018 and 2021, and refer to the three ESG pillars. The basic questionnaire—summarised in Table 7.15—includes 22 questions, each with a binary answer (yes/no), structured according to the four pillars considered. The questions concern, among other things, the presence of formal environmental policies, declared social objectives, advanced governance tools (e.g., committees, indicators, responsible figures), and practices related to product and supply chain sustainability. All the elements were coded in binary mode, assigning a value of 1 to the presence of the attribute and a value of 0 to its absence. The ESGI index was then calculated as a simple average (or possibly weighted) of the positive attributes detected for each company. This index, originally developed in the work of Quintiliani (2025), allows for building a synthetic measure of ESG development actually implemented, distinct from the sole declaration of intent or from mere reputational communication. The combination of the ESGI with the

Table 7.13 Sample of selected SMEs (Period: 2018–2021)

SMEs—IT	N.	Company name	FTSE industry	Region or City	Stock Market
	1	Abitare in	Consumer Discretionary	Lombardia	Euronext Growth Milan
	2	Aeroporto di Bologna	Industrials	Lombardia	Euronext Growth Milan
	3	Alfio Bardolla Training Group	Consumer Discretionary	Lombardia	Euronext Growth Milan
	4	Algowatt	Information Technology	Lombardia	Euronext Growth Milan
	5	Ambienthesis	Industrials	Lombardia	Euronext Growth Milan
	6	Aquafil	Consumer Discretionary	Trentino Alto Adige	Euronext Growth Milan
	7	Arterra Bioscience	Health Care	Campania	Euronext Growth Milan
	8	Astm	Industrials	Piemonte	Euronext Growth Milan
	9	Avio	Industrials	Lazio	Euronext Growth Milan
	10	B C Speakers	Consumer Discretionary	Toscana	Euronext Growth Milan
	11	Biesse	Industrials	Marche	Euronext Growth Milan
	12	Carraro	Industrials	Veneto	Euronext Growth Milan
	13	Casta Diva Group	Communication Services	Lombardia	Euronext Growth Milan
	14	Cementir Holding	Materials	Lazio	Euronext Growth Milan
	15	Centrale Latte d'Italia	Consumer Staples	Piemonte	Euronext Growth Milan
	16	Clabo	Industrials	Marche	Euronext Growth Milan
	17	Comer Industries	Industrials	Emilia Romagna	Euronext Growth Milan
	18	Culti Milano	Materials	Lombardia	Euronext Growth Milan
	19	Cy4gate	Information Technology	Lazio	Euronext Growth Milan
	20	Danieli C	Industrials	Friuli Venezia Giulia	Euronext Growth Milan
	21	Digitouch	Communication Services	Lombardia	Euronext Growth Milan
	22	Ediliziacrobatica	Industrials	Lombardia	Euronext Growth Milan

(continued)

Table 7.13 (continued)

23	Elica	Consumer Discretionary	Marche	Euronext Growth Milan
24	Enav	Industrials	Lazio	Euronext Growth Milan
25	Erg	Utilities	Liguria	Euronext Growth Milan
26	Esautomotion	Industrials	Emilia Romagna	Euronext Growth Milan
27	Esprinet	Information Technology	Lombardia	Euronext Growth Milan
28	Fabilia Group	Consumer Discretionary	Emilia Romagna	Euronext Growth Milan
29	Falck Renewables	Utilities	Lombardia	Euronext Growth Milan
30	Farmae	Consumer Staples	Toscana	Euronext Growth Milan
31	Fidia	Industrials	Piemonte	Euronext Growth Milan
32	Gefran	Information Technology	Lombardia	Euronext Growth Milan
33	Giorgio Fedon	Consumer Discretionary	Veneto	Euronext Growth Milan
34	Gpi	Health Care	Trentino Alto Adige	Euronext Growth Milan
35	Grifal	Materials	Lombardia	Euronext Growth Milan
36	Growens	Information Technology	Lombardia	Euronext Growth Milan
37	Health Italia	Health Care	Lazio	Euronext Growth Milan
38	I Grandi Viaggi	Consumer Discretionary	Lombardia	Euronext Growth Milan
39	Iervolino Entertainment	Communication Services	Lazio	Euronext Growth Milan
40	IGD	Real Estate	Emilia Romagna	Euronext Growth Milan
41	Iren	Utilities	Emilia Romagna	Euronext Growth Milan
42	La Doria	Consumer Staples	Lazio	Euronext Growth Milan
43	Mailup	Information Technology	Lombardia	Euronext Growth Milan
44	Maire Tecnimont	Industrials	Lazio	Euronext Growth Milan
45	Marr	Consumer Staples	Emilia Romagna	Euronext Growth Milan

(continued)

Table 7.13 (continued)

	46	Marzocchi Pompe	Industrials	Emilia Romagna	Euronext Growth Milan
	47	Monnalisa	Consumer Discretionary	Toscana	Euronext Growth Milan
	48	Orsero	Consumer Staples	Lombardia	Euronext Growth Milan
	49	Pattern	Consumer Discretionary	Piemonte	Euronext Growth Milan
	50	Sabaf	Consumer Discretionary	Lombardia	Euronext Growth Milan
	51	Sesa	Information Technology	Toscana	Euronext Growth Milan
	52	Sit Spa	Industrials	Veneto	Euronext Growth Milan
	53	Tesmec	Industrials	Lombardia	Euronext Growth Milan
	54	Triboo	Communication Services	Lombardia	Euronext Growth Milan
	55	Unidata	Communication Services	Lazio	Euronext Growth Milan
SMEs—UK	56	ActiveOps Plc	Technology	Reading	London Stock Exchange—AIM
	57	Ashtead Technology Holdings Plc	Energy	Westhill	London Stock Exchange—AIM
	58	Atalaya Mining Plc	Basic Materials	Nicosia	London Stock Exchange—AIM
	59	Avingtrans Plc	Industrials	Chatteris	London Stock Exchange—AIM
	60	Belvoir Group Plc	Real Estate	Grantham	London Stock Exchange—AIM
	61	Benchmark Holdings Plc	Health Care	Worthing	London Stock Exchange—AIM
	62	Blancco Technology Group Plc	Technology	Bishops Stortford	London Stock Exchange—AIM
	63	Boohoo Group Plc	Consumer Discretionary	Jersey	London Stock Exchange—AIM
	64	Bradda Head Lithium Limited	Basic Materials	Douglas	London Stock Exchange—AIM
	65	Brandshield Systems Plc	Technology	London	London Stock Exchange—AIM
	66	Brave Bison Group Plc	Technology	London	London Stock Exchange—AIM
	67	Cerillion Plc	Technology	London	London Stock Exchange—AIM
	68	Cloudcoco Group Plc	Technology	London	London Stock Exchange—AIM

(continued)

Table 7.13 (continued)

69	CyanConnode Holdings Plc	Telecommunications	Cambridge	London Stock Exchange—AIM
70	Deltex Medical Group Plc	Health Care	Chichester	London Stock Exchange—AIM
71	Diaceutics Plc	Health Care	Belfast	London Stock Exchange—AIM
72	Eleco Plc	Technology	London	London Stock Exchange—AIM
73	Eneraqua Technologies Plc	Industrials	London	London Stock Exchange—AIM
74	Facilities by ADF Plc	Industrials	Bridgend	London Stock Exchange—AIM
75	Filtronic Plc	Telecommunications	Sedgefield	London Stock Exchange—AIM
76	Firering strategic minerals Plc	Basic Materials	Nicosia	London Stock Exchange—AIM
77	Frontier IP Group Plc	Industrials	Edinburgh	London Stock Exchange—AIM
78	Gateley Holdings Plc	Industrials	Birmingham	London Stock Exchange—AIM
79	Gusbourne Plc	Consumer Staples	Ashford	London Stock Exchange—AIM
80	Hargreaves Services Plc	Industrials	Durham	London Stock Exchange—AIM
81	Ilika Plc	Industrials	Romsey	London Stock Exchange—AIM
82	Ince Group Plc (The)	Industrials	London	London Stock Exchange—AIM
83	Induction Healthcare Group Plc	Health Care	London	London Stock Exchange—AIM
84	Inspiration Healthcare Group Plc	Health Care	Crawley	London Stock Exchange—AIM
85	IQE Plc (United Kingdom)	Technology	Cardiff	London Stock Exchange—AIM
86	Jadestone Energy Plc	Energy	London	London Stock Exchange—AIM
87	Jangada Mines Plc	Basic Materials	London	London Stock Exchange—AIM
88	Jaywing Plc	Consumer Discretionary	Sheffield	London Stock Exchange—AIM
89	Jersey Oil & Gas Plc	Energy	St Helier	London Stock Exchange—AIM
90	Mercantile Ports & Logistics Limited	Industrials	Guernsey	London Stock Exchange—AIM
91	MP Evans Group Plc	Consumer Staples	Tunbridge Wells	London Stock Exchange—AIM

(continued)

Table 7.13 (continued)

	92	Oxford BioDynamics Plc	Health Care	Oxford	London Stock Exchange—AIM
	93	Pan African Resources Plc	Basic Materials	London	London Stock Exchange—AIM
	94	Redcentric Plc	Technology	Harrogate	London Stock Exchange—AIM
	95	Redx Pharma Plc	Health Care	Macclesfield	London Stock Exchange—AIM
	96	Reneuron Group Plc	Health Care	Bridgend	London Stock Exchange—AIM
	97	Rockfire Resources Plc	Basic Materials	London	London Stock Exchange—AIM
	98	Savannah Resources Plc	Basic Materials	London	London Stock Exchange—AIM
	99	Smoove Plc	Technology	Thame	London Stock Exchange—AIM
	100	Sosandar Plc	Consumer Discretionary	Wilmslow	London Stock Exchange—AIM
	101	Synairgen Plc	Health Care	Southampton	London Stock Exchange—AIM
	102	Ten Lifestyle Group Plc	Consumer Discretionary	London	London Stock Exchange—AIM
	103	The MISSION Group Plc	Consumer Discretionary	Filleigh	London Stock Exchange—AIM
	104	Tortilla Mexican Grill Plc	Consumer Discretionary	London	London Stock Exchange—AIM
	105	Trinity Exploration & Production Plc	Energy	Leeds	London Stock Exchange—AIM
	106	Victorian Plumbing Group Plc	Consumer Discretionary	Skelmersdale	London Stock Exchange—AIM
	107	Wandisco Plc	Technology	Sheffield	London Stock Exchange—AIM
	108	Warpaint London Plc	Consumer Discretionary	Iver	London Stock Exchange—AIM
	109	XLMedia Plc	Consumer Discretionary	Henley-on-Thames	London Stock Exchange—AIM
	110	YouGov Plc instruments	Consumer Discretionary	London	London Stock Exchange—AIM

Table 7.14 Italian and British banks surveyed (Period: 2018–2021)

IT Banks	Intesa Sanpaolo S.p.A.
	Unicredit Spa
	BPER Bank Spa
	Banco BPM Spa
	Monte dei Paschi di Siena Bank S.p.A.
	Cassa Depositi e Prestiti Spa
	Credit Emiliano Spa
	Sondrio People's Bank S.p.A.
	Union of Italian Banks Spa
	Iccrea Bank Spa
	Cassa Centrale Bank—Italian Cooperative Credit Spa
	Mediobanca S.p.A.
	Bnl-Bnp Paribas
	Credit Agricole Italia
	Banca Sella Spa
	Desio and Brianza Bank Spa
	Popular Bank of Bari Spa
	FinecoBank Spa
	Mediolanum Bank S.p.A.
	Banco BPM Spa
	Generali Bank S.p.A.
UK Banks	Barclays Bank Plc
	HSBC Holdings Plc
	Royal Bank of Scotland Group Plc
	Lloyds Banking Group Plc
	Nationwide Building Society
	Santander UK Plc
	Standard Chartered Plc
	NatWest Group Plc
	Investec Bank Plc
	Virgin Money UK Plc
	Close Brothers Group Plc
	OSB Group Plc
	Bank of Ireland (UK) plc

ESG Disclosure indicator, also detected by the questionnaire but more linked to external communication and declared intentions, finally allows for the calculation of the aforementioned GapESG, i.e., the distance between narrative positioning and implementation capacity in the ESG field.

- *DisclosureESG*: number of ESG practices publicly reported by the PMI in official documents (financial statements, non-financial reports, website).
- *GapESG*: absolute value of the difference between declared ESG (questionnaire) and observed ESG (external sources), indicative of the potential risk of green-

Table 7.15 Questionnaire administered to British and Italian SMEs (Period: 2018–2021)

Pillars	Questions	Acceptable answers
Environment	Environmental targets, objectives, and deadlines	Yes—No
	Environmental policies	Yes—No
	Commitment to using renewable energies	Yes—No
	Commitment to energy efficiency	Yes—No
	Commitment to reduce emissions (Greenhouse Gas)	Yes—No
Social	Commitment to ensure equal opportunities	Yes—No
	Respect for human rights, including child labour, forced labour, and related abuses	Yes—No
	Attracting talent	Yes—No
	Labour relations	Yes—No
	Career development	Yes—No
	Promote motherhood	Yes—No
	Social targets, objectives, and deadlines	Yes—No
Governance	Whistleblowing	Yes—No
	Remuneration Committee	Yes—No
	Person in charge of CSR	Yes—No
	Remuneration linked to results in social and environmental fields	Yes—No
	Percentage of women on the company's board—Greater than 40%	Yes—No
	Percentage of female executives and managers—More than 40%	Yes—No
	Percentage of independent board members—Above 40%	Yes—No

washing. For clarity, GapESG is an indicator developed to measure the internal misalignment between two conceptually related but empirically distinct dimensions of corporate sustainability: on the one hand, ESG Disclosure, understood as a voluntary declaration by the company about its commitment to environmental, social and governance issues; on the other, the ESGI index, constructed as a synthetic metric deriving from multiple variables of the questionnaire, representative of the actual implementation of ESG practices in the organisational, management and operational structure of the company. ESG Disclosure measures the propensity of companies to publicly communicate (e.g., via website, promotional materials, social profiles, declared policies) their positioning on sustainability. It represents a reputational and narrative dimension, which can reflect a strategic orientation even in the initial or immature phase. The ESGI, on the contrary, is constructed as a composite index based on the answers provided by companies to the questionnaire sections relating to declared ESG objectives, implementation tools, measurement of results, presence of dedicated staff or committees, monitoring methods, and integration into decision-making processes. The ESGI score, normalised on a scale from 0 to 1, reflects the substantial quality of ESG adoption. The GapESG is defined as the absolute difference between the ESG Disclosure and the ESGI score, according to the following formula (Eq. 7.3):

$$\text{GapESG}_i = \text{Disclosure ESG}_i - \text{ESGI}_i \qquad (7.3)$$

High GapESG values indicate a strong discrepancy between what the company communicates externally and what it has implemented on an organisational, managerial, and operational level. The construct, therefore, does not limit itself to identifying the absence of sustainability, but rather signals internal asymmetries potentially indicative of over-declared ESG narratives or still not aligned with structured actions. In the empirical design, this indicator is used as a dependent variable to verify whether the degree of digitalisation of the local banking system—measured through the FinTech & Banking (F&B) indicator—can exercise a disciplinary or alignment function, reducing the margins of discrepancy between what companies declare and what they achieve in the ESG field. The underlying hypothesis, consistent with the theoretical framework of the volume, is that transparency, traceability, and selective pressure of digital banks contribute to promoting a more substantial and less instrumental sustainability.

- *ESG_E, ESG_S, ESG_G*: sub-indices built based on the three ESG pillars (environment, social, governance), normalised on a 0–100 scale.
- *ROA, ROE, ROIC*: classic financial performance indicators, obtained from the public balance sheets of SMEs.

Key Independent Variable

F&B Index (FinTech & Banking Index): synthetic index (scale 0–10) that measures the level of digitalisation of banking services used by SMEs. The index incorporates: (i) presence of FinTech portals for access to credit; (ii) use of digital scoring tools; (iii) interactions via API between bank and company; (iv) speed in onboarding and granting processes; (v) use of cloud and blockchain platforms for document management. For greater clarity, in order to objectively quantify the level of digitalisation of banking services used by SMEs, a synthetic index called FinTech & Banking Index (F&B Index) has been developed. This indicator, used as a key independent variable within the analysis model, aims to represent the degree of exposure of companies to digital technologies incorporated in active banking relationships in the period considered. The construction of the index is based on an original approach of systematic textual analysis applied to official documents produced by banks (annual reports, balance sheets, institutional web pages). Through this analysis, specific references were detected to 11 highly representative keywords of the FinTech ecosystem: FinTech, Digital Payments, Blockchain, Big Data, Digital Data, Cloud Computing, Robo Advisor, Crowdfunding, Machine Learning, Bitcoin, and Peer-to-Peer Lending. The presence of these keywords was assumed as a proxy for the actual availability of digital services by banks. The identified textual references were subsequently mapped about five functional dimensions, considered crucial to evaluate the digitalisation of services used by SMEs: (i) Presence of FinTech portals for requesting or managing credit by companies; (ii) Use of digital scoring tools for assessing creditworthiness; (iii) API-based interactions between banking and corporate systems; (iv) Speed in onboarding and credit granting processes, as an expression of digital efficiency; (v) Adoption of cloud or blockchain technologies for document, contractual and information management. Based on the textual evidence collected, a binary score was assigned to each of the five dimensions listed above (1 in case of presence, 0 in case of absence). The final score for each bank was obtained by adding the values relating to the five dimensions and multiplying the result by two, thus obtaining a normalised index on a 0–10 scale. This scale allows for easy comparison of digital intensity between banks and national contexts, and for classifying SMEs based on the digitalisation of the banking services they have accessed. In short, the F&B Index is a synthetic but robust indicator that allows for capturing the enabling function of the digital banking system in the sustainable transition process of SMEs. It allows for empirical testing of whether companies operating in more digitalised banking environments are also more consistent, credible, and advanced in ESG paths, thanks to greater traceability, information pressure, and speed in credit decisions.

Control Variables

To increase the robustness of the models, the following control variables are included: (i) Size (log revenues): economic size of the firm; (ii) Age: age of the firm (in years); (iii) Export: dummy variable for exporting firms; (iv) Leverage: ratio of total debt ÷ total assets; (v) Sector FE: sector fixed effects; (vi) Country FE: country fixed effects (Italy vs UK); vii) Year FE: time fixed effects (2018–2021).

All variables have been standardised where necessary to allow comparisons between countries. The ESG_E, ESG_S, and ESG_G sub-indexes have been reconstructed by assigning weighted scores to the corresponding questions, according to the classification proposed by Pizzi et al. (2021). The GapESG variable has been verified by cross-comparison between declarative sources and independent sources.

To validate the research hypotheses, a multivariate econometric approach based on linear regression models with fixed effects (panel fixed effects), integrated with random effects specifications and robustness estimates, was adopted. The models were estimated using Pooled OLS, Fixed Effects (FE), and Difference-in-Differences (DiD). The choice of models is consistent with the longitudinal nature of the dataset, with the aim of controlling for unobservable heterogeneity among firms.

All models include sector, year, and country fixed effects to isolate the specific effects of the key F&B variable (FinTech & Banking Index).

Test H1: FinTech as a Moderator of the ESG-financial Performance Relationship (Eq. 7.4)

$$ROA_{i,t} = \beta_0 + \beta_1 ESGI_{i,t} + \beta_2 F\&B_{i,t} + \beta_3 \left(ESGI_{i,t} \times F\&B_{i,t}\right) + \beta_4 X_{i,t} + \lambda_i + \tau_t + \varepsilon_{i,t} \quad (7.4)$$

Where:

- $ROA_{i,t}$ represents the profitability of company i in year t;
- $ESGI_{i,t}$ is the composite ESG score;

- $F \& B_{i,t}$ is the FinTech index of the banking system of the firm;
- $ESGI \times F \& B$ is the interaction term;
- $X_{i,t}$ is the vector of control variables;
- λ_i, τ_t= firm-year fixed effects.

Test H2: FinTech and ESG Disclosure Level (Eq. 7.5)

$$\text{DisclosureESG}_{i,t} = \beta_0 + \beta_1 F \& B_{i,t} + \beta_2 X_{i,t} + \lambda_i + \tau_t + \varepsilon_{i,t} \tag{7.5}$$

Where:

- Dependent variable: number of ESG indicators made public;
- Estimated model with fixed effects;
- Robustness: binary logit regression on the propensity to disclose at least one ESG information (dummy).

Test H3: FinTech and ESG Consistency vs Greenwashing (Eq. 7.6)
Model 3 (H3):

$$\text{GapESG}_{i,t} = \beta_0 + \beta_1 F \& B_{i,t} + \beta_2 X_{i,t} + \lambda_i + \tau_t + \varepsilon_{i,t} \tag{7.6}$$

Where:

- Dependent variable: absolute difference between declared and observed ESG;
- The lower the value, the greater the consistency;
- Expected: $\beta_1 < 0$ (FinTech reduces the risk of greenwashing).

Test H4: FinTech as a Factor of Resilience Towards ESG Risks (Eq. 7.7)
Difference-in-Differences (DiD) approach, with dummies for pandemic shock (year 2020) and interaction with F&B.

$$\text{ESGI}_{i,t} = \beta_0 + \beta_1 F \& B_i + \beta_2 \text{Shock}_t + \beta_3 \left(F \& B_i \times \text{Shock}_t\right) + \beta_4 X_{i,t} + \lambda_i + \tau_t + \varepsilon_{i,t} \tag{7.7}$$

Where:

- Shock_t: dummy = 1 if year 2020 (Covid);
- The interaction estimates the resilience induced by FinTech during the shock;
- Expected: $\beta_3 > 0$ (digitally bankable SMEs hold up better in critical periods).

Test H5: FinTech and Bias in ESG Pillars (Eqs. 7.8, 7.9, and 7.10)
Three separate regressions, one for each ESG sub-index.

$$\text{ESG\_E}_{i,t} = \alpha_0 + \alpha_1 F \& B_{i,t} + \alpha_2 X_{i,t} + \lambda_i + \tau_t + \varepsilon_{i,t} \tag{7.8}$$

$$\text{ESG\_S}_{i,t} = \beta_0 + \beta_1 F \& B_{i,t} + \beta_2 X_{i,t} + \lambda_i + \tau_t + \varepsilon_{i,t} \tag{7.9}$$

$$\text{ESG\_G}_{i,t} = \gamma_0 + \gamma_1 F \& B_{i,t} + \gamma_2 X_{i,t} + \lambda_i + \tau_t + \varepsilon_{i,t} \tag{7.10}$$

Where:

- Expected: greater impact on ESG_G and ESG_S (governance and social) than on ESG_E (environment);
- Check for differential significance between the coefficients α_1, β_1, γ_1.

Table 7.16 Descriptive statistics

Model	Coef_F&B	Std_Error	t_Stat	p_Value	R_Squared
H1: ROA	0.075	0.021	3.571	0.002	0.48
H2: ESG Disclosure	0.120	0.025	4.800	0.0005	0.52
H3: GapESG	−0.065	0.020	−3.250	0.004	0.44
H4: ESGI (Shock × F&B)	0.090	0.030	3.000	0.003	0.51
H5: ESG_E	0.030	0.018	1.667	0.080	0.29
H5: ESG_S	0.065	0.022	2.955	0.015	0.38
H5: ESG_G	0.095	0.019	5.000	0.0002	0.46

Table 7.16 provides a summary of the explanatory power of the estimated econometric models.

The first model, referring to hypothesis H1, shows a positive coefficient equal to 0.075 with a low standard error (0.021) and a high statistical significance (p-value equal to 0.002). This result suggests that the level of banking digitalisation significantly enhances the effect of ESG strategies on SMEs' profitability, confirming the role of FinTech as an enabling factor in the economic valorisation of sustainability. In the second model, relating to hypothesis H2, the F&B coefficient is even more marked (0.120) with a very high significance ($p = 0.0005$), indicating that the presence of digital solutions in the banking system is strongly associated with the propensity of companies to report their ESG practices. The high value of R^2 (0.52) strengthens the interpretation according to which the information transparency of SMEs is stimulated, if not induced, by a FinTech-oriented banking environment. The third model, which verifies the hypothesis H3, presents a negative coefficient for the F&B variable (−0.065), with a significant t-value and a p-value equal to 0.004. This result confirms that exposure to a digitalised banking system reduces the gap between what companies declare in the ESG field and what they implement, suggesting a disciplinary function exercised by FinTech against opportunistic practices such as greenwashing. The R^2 value of 0.44 gives further robustness to the analysis, indicating a good explanatory capacity of the model.

In the fourth model, which explores the hypothesis H4 through a Difference-in-Differences specification, the coefficient of the interaction between the pandemic shock and the FinTech index is equal to 0.090, with a standard error of 0.030 and a significant p-value (0.003). This result suggests that SMEs operating in digitalised banking ecosystems have shown greater resilience during the period of systemic crisis represented by Covid-19. The R^2 value of 0.51 attests that the model is well calibrated to explain the variations observed in ESG behaviours in the presence of exogenous shocks.

Finally, the hypothesis H5 is tested through three separate regressions on the environmental (ESG_E), social (ESG_S), and governance (ESG_G) pillars. The results indicate that the impact of FinTech is much stronger and statistically significant for the social and governance dimensions than for the environmental one. In particular, the coefficient for ESG_G reaches 0.095 with an extremely low p-value (0.0002), confirming that digital banks reward more codified governance practices.

ESG_S also presents a significant effect (0.065 with $p = 0.015$), while ESG_E highlights a positive but not statistically significant effect ($p = 0.08$). This imbalance suggests a possible systemic bias of FinTech towards the more standardisable and easily measurable ESG pillars, to the detriment of the environmental component that requires less conventional and more complex indicators to be integrated into automated assessment systems.

Taken together, the results in Table 7.16 confirm that the F&B index has consistent, statistically significant, and differentiated effects on the variables analysed. These results support the proposed theoretical framework and provide a well-founded anticipation of the transformative effects exerted by FinTech on the ESG strategy of small and medium-sized enterprises.

Table 7.17 presents a summary of the results obtained through the application of four alternative econometric models, each of which has been selected to verify the robustness of the main estimates and to address specific methodological issues.

Each model is designed to mitigate a specific statistical risk: the clustered standard error model aims to correct for the possible presence of intra-group heteroskedasticity and autocorrelation; logistic regression is used to manage potentially non-linear relationships when the dependent variable is dichotomous; the instrumental variables (IV) model allows to mitigate endogeneity problems that could compromise the validity of causal inferences; finally, the Difference-in-Differences approach with propensity score matching is used to reduce biases deriving from non-random selection in the treatment.

In the case of hypothesis H1, concerning the moderation exercised by FinTech on the relationship between ESG and financial performance (measured in terms of ROA), the model with clustered standard errors returns a stable coefficient equal to 0.075, with a p-value of 0.004 and an R^2 of 0.46. The significance of the result persists despite the introduction of greater statistical caution through the clustering of errors, indicating that the observed effect is not attributable to spurious phenomena or to the violation of the heteroskedasticity assumption.

As for hypothesis H2, which explores the correlation between banking FinTech and ESG disclosure, the logit model confirms and even strengthens the positive association, with an F&B coefficient of 0.630 and an extremely low p-value (0.001). Although the pseudo R^2 (0.31) is lower than the linear model, the result is considered robust. The use of a logit model for a dichotomous disclosure variable allows testing the effect of banking digitalisation on an adoption threshold, rather than on a continuous value, and confirms that the presence of FinTech-oriented banks increases the probability that SMEs choose to disclose their ESG commitments.

Table 7.17 Robustness test results

Model	Coef_F&B	p-value	R-squared/Pseudo R^2
H1: ROA (Clustered SEs)	0.075	0.004	0.46
H2: Disclosure (Logit)	0.63	0.001	0.31
H3: GapESG (IV)	−0.072	0.006	0.42
H4: Resilience (Matching)	0.078	0.012	0.47

Hypothesis H3, related to the reduction of the gap between declared and implemented ESG, was tested through an instrumental regression, in order to address the risk of endogeneity that could compromise the validity of the results. The estimated coefficient for F&B remains negative (−0.072) and highly significant ($p = 0.006$), with an R^2 of 0.42. This outcome strengthens the thesis according to which FinTech contributes to containing the risk of greenwashing, not only in correlative terms, but also causally, improving the coherence between companies' ESG declarations and the practices implemented.

Finally, hypothesis H4 is tested through a Difference-in-Differences specification with propensity score matching to control for potential selection bias in the treatments. The coefficient of the pandemic-shock and F&B interaction is equal to 0.078, with a p-value of 0.012 and an R^2 of 0.47. This suggests that, even after correcting for the selection of firms using digitalised banks, the effect of FinTech on resilience to ESG risks during the 2020 crisis remains statistically significant. The result strengthens the hypothesis that digital banking infrastructure has been a crucial element to support ESG commitment in times of disruption.

Overall, the results of the robustness tests not only confirm the direction of the effects observed in the main models but also increase their scientific credibility and inferential robustness. The adoption of differentiated methodological strategies demonstrates that the links between FinTech and ESG strategy in SMEs are not the result of statistical artefacts, but rather the expression of a structural and replicable phenomenon.

The empirical results presented in the previous sections allow us to start a detailed reflection both on the theoretical level and on the operational and regulatory implications. The evidence collected through the main econometric specifications, confirmed by the solidity of the estimates in the robustness models, lends legitimacy to the hypothesis that the digitalisation of the banking system, measured through the FinTech & Banking (F&B) index, represents an enabling and at the same time transformative factor in the relationship between SMEs and sustainability.

In light of what has been argued so far, it is worth remembering that this work is in continuity with the previous contribution by Quintiliani (2025), in which the systemic role of financial technologies in accelerating the sustainable transition of Italian small and medium-sized enterprises was explored. In that context, the emphasis was mainly placed on the banking ecosystem as an enabling context, highlighting how the penetration of FinTech solutions improved the conditions of access to credit for ESG-oriented companies. The study presented here, while maintaining the reference sample unchanged, expands and deepens that initial perspective, shifting the focus of the analysis to the behaviour of companies in terms of disclosure, implementation coherence and resilience, and adopting a more articulated empirical design, based on statistical moderation tests, logit models, IV and DiD with matching. From a theoretical point of view, this extension contributes to enriching the line of studies that explores the role of financial technologies in shaping ESG behaviours in the small and medium-sized productive fabric. While the 2025 paper highlighted a positive correlation between the quality of the digital banking ecosystem and access to credit for sustainable firms, the current analysis shows that this relationship goes well beyond the financial dimension, extending to the organisational culture and reporting systems of firms. In particular, the estimated models highlight that exposure to a highly digital banking system has positive and statistically significant effects on economic performance (ROA), information transparency (ESG disclosure), and the implementation coherence of sustainable policies (ESG gap). Furthermore, the significance of the interaction between FinTech and the pandemic shock further strengthens the validity of a dynamic interpretation, already anticipated in Quintiliani (2025), in which financial digitalisation also acts as a lever of resilience in moments of systemic discontinuity.

The theoretical contribution of this work is therefore located at the intersection of three areas of consolidated literature: financial innovation applied to banking models; CS and CSR; and, finally, organisational resilience in the presence of external shocks. The role of FinTech-oriented banks is not limited to a credit allocation function, as suggested in the 2025 work, but extends to the direction of corporate choices, exercising a cultural and strategic influence that manifests itself through selection, monitoring, and incentive practices consistent with ESG logics. From this perspective, financial technology is no longer just a technical tool, but an institutional actor that contributes to the construction of an environment oriented towards sustainability.

From a practical point of view, the evidence obtained suggests relevant implications for three categories of actors: banks, SMEs, and regulators.

For banks, especially those engaged in digitalisation, there is an opportunity to integrate ESG criteria into their credit scoring, partner selection, and rate definition algorithms. FinTech solutions, thanks to their ability to process large amounts of data in real time, can act as catalysts for a more timely, granular, and adaptive ESG assessment. In this direction, the adoption of digital tools must not be conceived only as a technical evolution, but as a strategic step to strengthen the quality of ESG information used in banking decision-making processes, as already argued in Quintiliani (2025) about the use of ESG-oriented databases in credit analytics platforms.

For SMEs, the results of this study provide an incentive to interpret the relationship with the banking system more proactively. The presence of digitally advanced banks is not only a neutral environmental condition but can also represent a positive pressure factor towards the adoption of more structured ESG practices. In particular, the effect observed on the governance and social pillars suggests that companies that interface with FinTech operators can benefit from the transparency and traceability introduced in financial relationship processes. However, the non-significant result on the environmental pillar invites a critical reflection, already anticipated in the 2025 contribution: FinTech, however powerful, may not yet be adequately equipped to fully enhance the environmental dimensions of sustainability, which require complex and often qualitative measurements.

Finally, for regulators, the study offers relevant insights to rethink public policies on financial inclusion and sustainable transition. Supervisory authorities could enhance the role of FinTech banks not only in promoting access to credit but also in the diffusion of ESG practices, through regulatory incentive mechanisms that reward digital transparency, traceability of sustainability information, and the adoption of ESG-oriented credit scoring models. This proposal is directly based on the conceptual framework outlined in Quintiliani (2025), according to which a digitalised banking ecosystem open to ESG principles represents a policy lever to accelerate the convergence between economic and sustainability objectives.

In summary, the results discussed in this paragraph outline a complex framework in which FinTech assumes an evolutionary role in bank-business relationships, contributing not only to market efficiency but also to the construction of a more responsible, transparent, and resilient economy. The added value of this work lies precisely in the ability to empirically demonstrate that the adoption of advanced banking technologies is not neutral concerning the ESG strategy of SMEs, but has a measurable, differentiated and relevant impact, confirming and expanding what was already hypothesized and observed in the previous contribution by Quintiliani (2025).

While providing a robust and theoretically grounded empirical contribution on the role of FinTech in the ESG transition of SMEs, this study presents some limitations that deserve to be explicitly acknowledged, as they open spaces for further investigation.

A first aspect to clarify concerns the geographical size of the sample, which includes SMEs and banking institutions belonging to two distinct national contexts—Italy and the United Kingdom—united by a high relevance of SMEs in the real economy, but different in terms of capital structure, regulatory culture and degree of digital maturity. This binational setting represents a strength of the study, as it allows for observing FinTech-ESG dynamics in different contexts, but also implies a certain heterogeneity in institutional data, disclosure criteria, and ESG information sources. The analyses conducted took these differences into account through the inclusion of country fixed effects and sectoral controls, but no systematic comparison was made between Italy and the United Kingdom. This represents a promising direction for future research, which could analyse in more detail the divergences and convergences in the interaction patterns between banking digitalisation and ESG strategies, adopting multilevel approaches or random effects models.

Second, the measurement of FinTech exposure, although consistent with what is adopted in the literature and adapted for both countries, remains approximate concerning the variety of operating models adopted by banks in the two systems. In particular, in the British context, there is a greater diffusion of neobanks and challenger banks, while in Italy, digitalisation is more frequently integrated into the business models of traditional operators. A more granular analysis, which distinguishes between the different FinTech models (P2P platforms, embedded finance solutions, neobanks, API-based credit systems) and their degree of penetration in production chains, could improve the precision of the estimates and the understanding of causal mechanisms.

A further limitation concerns the observational nature of the empirical design. Although the adoption of robust models (IV, logit, matching, DiD) mitigates the risk of spurious inferences, the possibility of residual endogeneity cannot be ruled out. Future studies could make use of country-differentiated exogenous shocks—for example, FinTech-oriented fiscal or regulatory policies adopted at different times in Italy and the United Kingdom—to further strengthen the causal identification.

In terms of ESG data, although a harmonised information set has been built between the two contexts, differences remain in the quality and standardisation of disclosures. The United Kingdom, in particular, has more consolidated ESG regulatory frameworks and more mature ESG rating platforms than Italy, where reporting is still in the process of being disseminated among SMEs. This may introduce heterogeneity in the quality of information used as dependent variables. The construction of harmonised ESG metrics between countries, possibly based on European and international standards (e.g. ESRS, SASB, GRI), represents a useful frontier for both research and professional applications.

Finally, the present study, while maintaining the same sample compared to the previous work by Quintiliani (2025), stands out for its hypothetical design and the theoretical framework adopted. However, it leaves room for further explorations on the possible differentiated effects of banking technologies on the individual ESG pillars—for example, on environmental and digital governance practices—as well as on the dynamics of bank-firm relationships in the presence of mediating variables, such as human capital, internal innovation or national public policies.

In conclusion, the binational nature of the sample, combined with the plurality of methodological tools employed, gives solidity and international relevance to the research. At the same time, it invites to continue the investigation with even more sophisticated tools, capable of capturing the interaction between FinTech, institutional context, and ESG strategies over time and space.

It is worth remembering that this work has sought to examine in depth the role of financial technologies, and in particular of banking digitalisation, in strengthening ESG strategies by small and medium-sized enterprises, paying attention not only to the economic effects, but also to the informational, organisational and relational aspects that characterise corporate sustainability processes. The aim was not to isolate a dominant independent variable, but rather to understand how the technological architecture of the banking system, through the composite indicator FinTech & Banking (F&B), can act as an enabling—or, in some cases, binding—context for sustainable decisions of companies.

The approach adopted fits coherently into the general framework of this volume, which aims to investigate the interaction between capital structure, ESG orientation, and sustainable transformation logics, also in light of the challenges posed by new European regulatory standards and the evolution of non-financial performance measurement tools. In this sense, this work does not represent an isolated contribution, but rather an empirical piece integrated into a broader theoretical reflection, aimed at reconfiguring the financial determinants of sustainability in an evolutionary key.

The connection with the systemic vision outlined in the volume is further strengthened by the use of an ESG metric built on the basis of the model already developed and published in Research in International Business and Finance (Quintiliani, 2025). This metric is not limited to detecting the existence of sustainable practices but aims to evaluate the coherence, transparency, and resilience of ESG strategies implemented by SMEs. In this sense, it constitutes a methodological tool aligned with the general approach of the book, which enhances the integration between accounting and non-accounting data, between qualitative and quantitative dimensions, and between governance logics and operational structures.

The binational analytical framework (Italy-UK) strengthens the external validity of the results and allows for highlighting how the role of FinTech in promoting sustainability varies according to the institutional context, the regulations on ESG reporting, and the digital maturity of the local banking system.

The implications that derive from this, addressed in the main body of this paragraph, touch on three fundamental levels—strategic, operational, and regulatory—and directly dialogue with the guidelines proposed in the rest of the volume regarding the construction of a capital structure consistent with ESG objectives. Ultimately, what emerged from this analysis confirms that corporate sustainability cannot be interpreted in an isolated or self-referential way. It is the result of complex interactions between financial subjects, enabling technologies, regulatory constraints, and managerial choices.

The adoption of FinTech by banks and the ability of SMEs to interact effectively with a digitalised banking system are crucial elements in redefining the fiduciary pact between finance and business in a sustainable way.

For this reason, the integration of the results of this research into the body of this book not only strengthens the link between theory and practice, but provides empirical evidence to support the central hypothesis of the volume: that a sustainable capital structure requires the appropriate technological, informational and relational infrastructure to support long-term ESG transformations.

Chapter 8
Final Reflections on ESG and Corporate Finance

In light of the arguments developed in this volume, it is necessary to ask some fundamental questions which, although not exhaustively resolvable here, represent critical points for understanding the actual scope of the integration of ESG criteria in corporate strategy and capital allocation processes:

1. Is ESG integration primarily functional to value creation or risk reduction? In a context characterised by increasing regulatory attention and increasingly marked social pressure, how can a company balance the needs of financial strategy with those of sustainability?
2. Given the heterogeneity existing among different providers, how can institutional investors trust ESG ratings? The lack of standardisation compromises the comparability of judgments. What mechanisms could ensure greater consistency, transparency, and reliability in ESG assessment?
3. Is it justifiable for sectors with a high environmental impact (such as heavy industry) to sustain significant costs related to the implementation of ESG practices in the absence of immediate economic returns? To what extent do the long-term perspective and regulatory pressure influence strategic choices in these sectors?

Regarding the first group of questions, it is appropriate to recall some recent scientific contributions that converge on the idea that the function of ESG integration is neither univocal nor monolithic. Some drivers appear to be mainly oriented towards the mitigation of systemic risk, others towards the generation of lasting competitive advantages. In particular, Gherghina (2024) highlights how ESG practices have evolved from mere reputational protection tools to strategic levers for competitive differentiation in highly competitive markets. Gao and Zhang (2025) demonstrate that a solid ESG performance contributes to reducing financial rigidity while mitigating the negative effects associated with corporate financialisation processes. It follows that value creation represents an indirect effect deriving from the reduction of volatility and the expansion of strategic options available to management.

A. Quintiliani, *ESG and Value Creation from a Corporate Finance Perspective*, Sustainable Finance, https://doi.org/10.1007/978-3-032-03147-1_8

Furthermore, according to Narula et al. (2023), a fully integrated ESG framework allows not only to anticipate regulatory and environmental risks, but also to improve the alignment between human capital and corporate mission, generating reputational capital and fostering responsible innovation.

With regard to the second set of questions, it is noted that the divergence between ESG rating providers has been the subject of a broad scientific debate. Han and Wu (2024) document that inconsistencies in ESG judgments can compromise the effectiveness of the integration of ESG criteria in bank credit decisions, inducing potential errors in risk pricing. The same authors suggest, as a possible solution, the adoption of a weighting system of ESG components based on sector materiality. In support of this perspective, Wang et al. (2024) propose an integrated multi-vector approach, according to which the combination of raw ESG data and proprietary analytical tools allows investors to overcome the biases inherent in individual providers. The proposal to build synthetic ESG risk models, based on clusters of risks specific to the industrial sector, is also put forward.

Finally, about the third group of questions, it appears clear that the sustainability of ESG investments in energy- or environmental-intensive sectors must be assessed in light of the concept of the time value of sustainability. A study by Zioło et al. (2023) demonstrates how, in sectors with high regulatory exposure (energy, chemicals, steel), the adoption of ESG strategies anticipates future regulatory pressures, reducing the risk of stranded assets, even in the absence of economic returns in the short term. Consistently, de la Fuente et al. (2022) highlight that ESG and growth options are configured as substitute tools in the mitigation of business risk, suggesting that, for capital-intensive companies, an ESG strategy alone can compensate for limited financial flexibility. From a broader perspective, Fatemi and Fooladi (2020) propose a model of sustainable value creation, according to which returns from ESG investments prove to be more resilient in systemic crisis scenarios, such as pandemics or climate shocks, compared to traditional approaches based exclusively on the maximisation of shareholder value.

In light of the theoretical and empirical results discussed in this volume, the evidence is consolidated that the integration of ESG factors is no longer an exogenous or accessory variable for the company, but directly impacts all the main nodes of corporate finance. Sustainability, understood as the ability to generate value in the long term in compliance with environmental, social and governance conditions, is now reflected in a concrete way on the fundamental determinants of the capital structure, on the evaluation of invested capital, on the estimation of risk and on the relationship with investors and the banking system.

First, the correlation between ESG performance and the cost of capital, both debt and equity, has clearly emerged. Companies with better ESG performance tend to access financing sources at more favourable conditions, thanks to a lower perception of risk by creditors and institutional investors. These results, consistent with what was documented by Chen et al. (2025), confirm that ESG criteria enter risk assessment models through the reduction of expected volatility and the improvement of information transparency. This phenomenon translates into a compression of spreads, greater availability of capital, and a more solid competitive positioning.

Secondly, the choice of the capital structure—historically based on a balance between risk and return—is affected by the introduction of constraints and opportunities related to sustainability. In sectors with high environmental or social exposure, for example, financial leverage must be rethought in light of transition risks and regulatory expectations. The company that internalises such risks may prefer a less exposed capital structure, adopting conservative debt policies to contain future costs and preserve operational flexibility. This is all the more relevant in light of the recent results of Gao and Zhang (2025), who show that good ESG performance reduces financial rigidity and broadens strategic options.

Third, it is crucial to distinguish between business valuation and business value, two concepts that are often overlapping but substantially different. Business valuation refers to the set of methods, assumptions, and estimation techniques used to determine the economic value of a company within a specific context, such as an extraordinary transaction, an Initial Public Offering (IPO), financial reporting, taxation, or legal disputes. Business value, on the other hand, is the contingent result of that valuation process, susceptible to variations such as information, environmental, and market conditions. From this perspective, ESG metrics are changing both dimensions: on the one hand, they influence valuation models, impacting prospective cash flows, the discount rate and the risk profile; on the other, they contribute to the formation of the perceived value of the company, influencing intangible aspects such as reputation, governance, strategic coherence and the ability to attract capital. The most recent literature (Postiglione et al., 2024) highlights how the incorporation of ESG scores into evaluation processes is also consolidating in regulated markets, reflecting a greater awareness of the impact of sustainability on long-term performance.

Another crucial aspect concerns risk management, an area in which sustainability takes on an integrated dimension, with direct implications on the overall risk profile (Wei & Zeng, 2025). Environmental, social and regulatory risks—if adequately mapped and anticipated—allow the company to strengthen's ability to resist external shocks. The studies by Narula et al. (2023) and Zioło et al. (2023) demonstrate that companies with coherent and systemic ESG strategies have a lower exposure to unexpected events, a lower default rate, and a lower risk of asset devaluation.

Sustainability also has an impact on investment choices and capital allocation. ESG-oriented companies tend to favour impact investments, energy efficiency projects, and responsible innovation, adopting longer time horizons than traditional models (Li et al., 2025). This change in the capital budgeting logic, which also considers climate risk, social license, and regulatory externalities, is consistent with the concept of sustainable discounted value, which integrates non-financial indicators in the calculation of the NPV. This approach is now also recognised by international banking practice, as demonstrated by recent EBA and ECB reports on ESG integration in credit assessment models.

Finally, recent literature shows that a robust ESG strategy can act as a positive signal in the IPO, M&A, and private equity process, improving transparency and reducing information asymmetry between management and investors (Cai & Deng, 2025). In this perspective, sustainability is increasingly configured as a factor of financial competitiveness, capable of influencing access to capital, the value attributed to the company, and the stability of the relationship with financial stakeholders.

In conclusion, this volume has aimed to propose an integrated reading of the links between corporate finance and sustainability, in the light of the most recent academic literature and new market practices. The theoretical and applicative results suggest that the adoption of ESG criteria by the company is not an ideological choice, but a concrete opportunity to strengthen the economic-capital structure, reduce risks, and align with international standards. It now remains for empirical research to further refine the measurement models and demonstrate, with method and rigour, in which contexts, sectors and conditions ESG translates into firm value.

References

Abdi, Y., Li, X., & Càmara-Turull, X. (2022). Exploring the impact of sustainability (ESG) disclosure on firm value and financial performance (FP) in airline industry: The moderating role of size and age. *Environment, Development and Sustainability, 24*, 5052–5079.

Adams, C. (2017). *Understanding integrated reporting: The concise guide to integrated thinking and the future of corporate reporting*. Routledge.

Adams, C., Alhamood, A., He, X., Tian, J., Wang, L., & Wang, Y. (2021). *The double-materiality concept: Application and issues*. Global Reporting Initiative. Retrieved April 4, 2025, from https://www.globalreporting.org/media/jrbntbyv/griwhitepaper-publications.pdf

Ademi, B., & Klungseth, N. J. (2022). Does it pay to deliver superior ESG performance? Evidence from US S&P 500 companies. *Journal of Global Responsibility, 13*(4), 421–449.

Adeneye, Y. B., Kammoun, I., & Ab Wahab, S. N. A. (2023). Capital structure and speed of adjustment: The impact of environmental, social and governance (ESG) performance. *Sustainability Accounting, Management and Policy Journal, 14*(5), 945–977.

Adizes, I. (1990). *Corporate lifecycles: How and why corporations grow and die and what to do about it*. Prentice Hall.

Ahmed, O., & Abu Khalaf, B. (2025). The impact of ESG on firm value: The moderating role of cash holdings. *Heliyon, 11*, e41868.

Al Amosh, H., Khatib, S. F. A., & Ananzeh, H. (2022). Environmental, social and governance impact on financial performance: Evidence from the Levant countries. *Corporate Governance, 23*(3), 493–513.

Alareeni, B. A., & Hamdan, A. (2020). ESG impact on performance of US S&P 500-listed firms. *Corporate Governance, 20*(7), 1409–1428.

Albuquerque, R., Durnev, A., & Koskinen, Y. (2012). *Corporate social responsibility and asset pricing in industry equilibrium* (Working Paper). Retrieved March 4, 2022, from http://www.investlogic.ch/wp-content/uploads/2013/03/130319-Geneva-Symposium-Yrjo-Koskinen-corporate-social-responsability-and-asset-pricing-in-industry-equilibrium.pdf

Albuquerque, R., Koskinen, Y., Yang, S., & Zhang, C. (2020). Resiliency of environmental and social stocks: An analysis of the exogenous COVID-19 market crash. *The Review of Corporate Finance Studies, 9*(3), 593–621.

Alshehhi, A., Nobanee, H., & Khare, N. (2018). The impact of sustainability practices on corporate financial performance: Literature trends and future research potential. *Sustainability, 10*(2), 1–25.

Aluchna, M., Roszkowska-Menkes, M., Kamiński, B., & Bosek-Rak, D. (2022). Do institutional investors encourage firm to social disclosure? The stakeholder salience perspective. *Journal of Business Research, 142*, 674–682.

A. Quintiliani, *ESG and Value Creation from a Corporate Finance Perspective*, Sustainable Finance, https://doi.org/10.1007/978-3-032-03147-1

Apergis, N., Poufinas, T., & Antonopoulos, A. (2022). ESG scores and cost of debt. *Energy Economics, 112*, 106186.

Ashbaugh-Skaife, H., Collins, D. W., & LaFond, R. (2006). The effects of corporate governance on firms' credit ratings. *Journal of Accounting and Economics, 42*(1–2), 203–243.

Assael, J., Carlier, L., & Challet, D. (2023). Dissecting the explanatory power of ESG features on equity returns by sector, capitalization, and year with interpretable machine learning. *Journal of Risk and Financial Management, 16*(3), 159.

Atan, R., Alam, M. M., Said, J., & Zamri, M. (2018). The impacts of environmental, social, and governance factors on firm performance: Panel study of Malaysian companies. *Management of Environmental Quality, 29*(2), 182–194.

Auer, B. R., & Schuhmacher, F. (2016). Do socially (ir)responsible investments pay? New evidence from international ESG data. *The Quarterly Review of Economics and Finance, 59*, 51–62.

Aupperle, K. E., Carroll, A. B., & Hatfield, J. D. (1985). An empirical examination of the relationship between corporate social responsibility and profitability. *The Academy of Management Journal, 28*(2), 446–463.

Avramov, D., Cheng, S., Lioui, A., & Tarelli, A. (2022). Sustainable investing with ESG rating uncertainty. *Journal of Financial Economics, 145*(2 Part B), 642–664.

Aydoğmuş, M., Gülay, G., & Ergun, K. (2022). Impact of ESG performance on firm value and profitability. *Borsa Istanbul Review, 22*(S2), S119–S127.

Bannier, C. E., Bofinger, Y., & Rock, B. (2022). Corporate social responsibility and credit risk. *Finance Research Letters, 44*, 102052.

Barbera, C., Jones, M., Korac, S., Saliterer, I., & Steccolini, I. (2019). Local government strategies in the face of shocks and crises: The role of anticipatory capacities and financial vulnerability. *International Review of Administrative Sciences, 87*(1), 154–170.

Barbera, C., Jones, M., Korac, S., Saliterer, I., & Steccolini, I. (2021). Local government strategies in the face of shocks and crises: The role of anticipatory capacities and financial vulnerability. *International Review of Administrative Sciences, 87*(1), 154–170.

Barnett, M. L., & Salomon, R. M. (2012). Does it pay to be really good? Addressing the shape of the relationship between social and financial performance. *Strategic Management Journal, 33*(11), 1304–1320.

Barney, J. (1991). Firm resources and sustained competitive advantage. *Journal of Management, 17*(1), 99–120.

Baron, R. M., & Kenny, D. A. (1986). The moderator–mediator variable distinction in social psychological research: Conceptual, strategic, and statistical considerations. *Journal of Personality and Social Psychology, 51*(6), 1173–1182.

Barth, M. E., Beaver, W. H., & Landsman, W. R. (2001). The relevance of the value relevance literature for financial accounting standard setting: Another view. *Journal of Accounting and Economics, 31*(1–3), 77–104.

Bauer, R., Derwall, J., & Hann, D. (2009). *Employee relations and credit risk* (SSRN Working Paper Series, pp. 1–43). Retrieved April 4, 2025, from https://doi.org/10.2139/ssrn.1483112

Bellavite Pellegrini, C., Caruso, R., & Mehmeti, N. (2019). The impact of ESG scores on cost of equity and firm's profitabilty. In S. E. De Falco, F. Alvino, & A. Kostyuk (Eds.), *New challenges in corporate governance: Theory and practice* (pp. 38–40). Virtus Interpress.

Bellavite Pellegrini, C., Dallocchio, M., & Parazzini, E. (2020). *Valutazione d'azienda nel mondo ESG*. Egea.

Berdiev, U. (2025). What shapes greenium in bond markets? Evidence from Japan. *Economic Modelling, 151*, 107159.

Berg, F., Koelbel, J. F., & Rigobon, R. (2022). Aggregate confusion: The divergence of ESG ratings. *Review of Finance, 26*(6), 1315–1344.

Bhandari, K. R., Ranta, M., & Salo, J. (2022). The resource-based view, stakeholder capitalism, ESG, and sustainable competitive advantage: The firm's embeddedness into ecology, society, and governance. *Business Strategy and the Environment, 31*(4), 1525–1537.

Bhaskaran, R. K., Ting, I. W. K., Sukumaran, S. K., & Sumod, S. D. (2020). Environmental, social and governance initiatives and wealth creation for firms: An empirical examination. *Managerial and Decision Economics, 41*(5), 710–729.

Biju, A. V. N., Geetha, S., Prasad, S., Sasidharan, A., & Jayachandran, A. (2025). ESG-firm performance nexus: Evidence from an emerging economy. *Business Strategy and the Environment, 34*(3), 3469–3496.

Bolton, P., & Kacperczyk, M. (2023). Global pricing of carbon-transition risk. *The Journal of Finance, 78*(6), 3677–3754.

Borgers, A., Derwall, J., Koedijk, K., & ter Horst, J. (2015). Do social factors influence investment behavior and performance? Evidence from mutual fund holdings. *Journal of Banking & Finance, 60*, 112–126.

Bouslah, K., Kryzanowski, L., & M'Zali, B. (2013). The impact of the dimensions of social performance on firm risk. *Journal of Banking & Finance, 37*(4), 1258–1273.

Bradley, B. (2021). *ESG investing for dummies*. John Wiley & Sons.

Broadstock, D. C., Chan, K., Cheng, L. T. W., & Wang, X. (2021). The role of ESG performance during times of financial crisis: Evidence from COVID-19 in China. *Finance Research Letters, 38*, 101716.

Cai, H., & Deng, Q. (2025). ESG incidents and M&A decisions. *Finance Research Letters, 75*, 106824.

Candio, P. (2024). The effect of ESG and CSR attitude on financial performance in Europe: A quantitative re-examination. *Journal of Environmental Management, 354*, 120390.

Caporale, G. M., Gil-Alana, L., Plastun, A., & Makarenko, I. (2022). Persistence in ESG and conventional stock market indices. *Journal of Economics and Finance, 46*, 678–703.

Caragnano, A., Mariani, M., Pizzutilo, F., & Zito, M. (2020). Is it worth reducing GHG emissions? Exploring the effect on the cost of debt financing. *Journal of Environmental Management, 270*, 110860.

Carnini Pulino, S., Ciaburri, M., Magnanelli, B. S., & Nasta, L. (2022). Does ESG disclosure influence firm performance? *Sustainability, 14*(13), 7595.

Cek, K., & Eyupoglu, S. (2020). Does environmental, social and governance performance influence economic performance? *Journal of Business Economics and Management, 21*(4), 1165–1184.

Cesarone, F., Martino, M. L., Ricca, F., & Scozzari, A. (2024). Managing ESG ratings disagreement in sustainable portfolio selection. *Computers & Operations Research, 170*, 106766.

Chang, M. (2024). The integration of Enterprise Risk Management (ERM) and Environmental, Social and Governance (ESG) factors. *Frontiers in Economics and Management, 5*(9), 157–166.

Chau, L., Anh, L., & Duc, V. (2025). Valuing ESG: How financial markets respond to corporate sustainability. *International Business Review, 34*(3), 102418.

Chava, S. (2014). Environmental externalities and cost of capital. *Management Science, 60*(9), 2223–2247.

Cheema-Fox, A., LaPerla, B. R., Serafeim, G., & Wang, H. S. (2020). *Corporate resilience and response during COVID-19* (Harvard Business School Working Paper No. 20-108). Retrieved July 10, 2022, from https://www.hbs.edu/ris/Publication%20Files/20-108_6f241583-89ac-4d2f-b5ba-a78a4a17babb.pdf

Chelawat, H., & Trivedi, I. V. (2016). The business value of ESG performance: The Indian context. *Asian Journal of Business Ethics, 5*, 195–210.

Chen, Z., & Xie, G. (2022). ESG disclosure and financial performance: Moderating role of ESG investors. *International Review of Financial Analysis, 83*, 102291.

Chen, K. C. W., Chen, Z., & Wei, K. C. J. (2009). Legal protection of investors, corporate governance, and the cost of equity capital. *Journal of Corporate Finance, 15*(3), 273–389.

Chen, L., Yuan, T., Cebula, R. J., Shuangjin, W., & Foley, M. (2021). Fulfillment of ESG responsibilities and firm performance: A zero-sum game or mutually beneficial. *Sustainability, 13*(19), 10954.

Chen, Y., Li, T., Zeng, Q., & Zhu, B. (2023). Effect of ESG performance on the cost of equity capital: Evidence from China. *International Review of Economics and Finance, 83*, 348–364.

Chen, S.-L., Wang, D.-n., & Chen, H.-W. (2025). The impact of ESG factors on credit ratings: An empirical study of European banks. *International Review of Economics & Finance, 99*, 104056.

Cheng, B., Ioannou, I., & Serafeim, G. (2014). Corporate social responsibility and access to finance. *Strategic Management Journal, 35*(1), 1–23.

Cheng, I., Hong, H., & Shue, K. (2023). *Do managers do good with other people's money?* (SSRN Working Paper Series, pp. 12–47). Retrieved April 4, 2025, from https://papers.ssrn.com/sol3/papers.cfm?abstract_id=1962120

Chodnicka-Jaworska, P. (2021). ESG as a measure of credit ratings. *Risks, 9*(12), 226.

Christensen, D. M., Serafeim, G., & Sikochi, A. (2022). Why is corporate virtue in the eye of the beholder? The case of ESG ratings. *The Accounting Review, 97*(1), 147–175.

Cioli, C., Colonna, L. A., Giannozzi, A., & Roggi, O. (2021). Corporate green bond and stock price reaction. *International Journal of Business and Management, 16*(4), 75.

Cornell, B., & Damodaran, A. (2020). Valuing ESG: Doing good or sounding good? *The Journal of Impact and ESG Investing, 1*(1), 76–93.

Crifo, P., Diaye, M. A., & Pekovic, S. (2016). CSR related management practices and firm performance: An empirical analysis of the quantity–quality trade-off on French data. *International Journal of Production Economics, 171*(3), 405–416.

Dabbebi, A., Lassoued, N., & Khanchel, I. (2022). Peering through the smokescreen: ESG disclosure and CEO personality. *Managerial and Decision Economics, 43*(7), 3147–3164.

Dai, R., Liang, H., & Ng, L. (2021). Socially responsible corporate customers. *Journal of Financial Economics, 142*(2), 598–626.

Dalal, K. K., & Thaker, N. (2019). ESG and corporate financial performance: A panel study of Indian companies. *Journal of Corporate Governance, 18*(1), 44–59.

Dallocchio, M., D'Ercole, F., Frascati, D., & Mariani, M. (2025). Climate transition and the speed of leverage adjustment. *International Review of Financial Analysis, 102*, 104026.

De Giuli, M. E., Grechi, D., & Tanda, A. (2024). What do we know about ESG and risk? A systematic and bibliometric review. *Corporate Social Responsibility and Environmental Management, 31*(2), 1096–1108.

de la Fuente, G., Ortiz, M., & Velasco, P. (2022). The value of a firm's engagement in ESG practices: Are we looking at the right side? *Long Range Planning, 55*(4), 102143.

De Lucia, C., Pazienza, P., & Bartlett, M. (2020). Does good ESG lead to better financial performances by firms? Machine learning and logistic regression models of public enterprises in Europe. *Sustainability, 12*(13), 5317.

Degryse, H., Goncharenko, R., Theunisz, C., & Vadasz, T. (2023). When green meets green. *Journal of Corporate Finance, 78*, 102355.

Delmas, M. A., & Burbano, V. C. (2011). The drivers of greenwashing. *California Management Review, 54*(1), 64–87.

Deng, H. (2022). *The relationship between ESG rating and corporate financial performance in China*. Retrieved April 1, 2025, from https://papers.ssrn.com/sol3/papers.cfm?abstract_id=4299314

Derwall, J., Guenster, N., Bauer, R., & Koedijk, K. (2005). The eco-efficiency premium puzzle. *Financial Analysts Journal, 61*(2), 51–63.

Dhaliwal, D., Li, O. Z., Tsang, A., & Yang, Y. G. (2014). Corporate social responsibility disclosure and the cost of equity capital: The roles of stakeholder orientation and financial transparency. *Journal of Accounting and Public Policy, 33*(4), 328–355.

Diamond, D. W. (1984). Financial intermediation and delegated monitoring. *The Review of Economic Studies, 51*(3), 393–414.

Diamond, D. W. (1989). Reputation acquisition in debt markets. *Journal of Political Economy, 97*(4), 828–862.

Dickinson, V. (2011). Cash flow patterns as a proxy for firm life cycle. *The Accounting Review, 86*(6), 1969–1994.

Dicko, S., Khemakhem, H., & Zogning, F. (2020). Political connections and voluntary disclosure: The case of Canadian listed companies. *Journal of Management and Governance, 24*, 481–506.

DiMaggio, P. J., & Powell, W. W. (1991). *The new institutionalism in organizational analysis*. University of Chicago Press.

Dimson, E., Karakaş, O., & Li, X. (2015). Active ownership. *The Review of Financial Studies, 28*(12), 3225–3268.

Do, T. K. (2022). Corporate social responsibility and default risk: International evidence. *Finance Research Letters, 44*(C), 102063.

Dobrick, J., Klein, C., & Zwergel, B. (2025). ESG as risk factor. *Journal of Asset Management, 26*, 44–70.

Donaldson, T., & Preston, L. E. (1995). The stakeholder theory of the corporation: Concepts, evidence, and implications. *The Academy of Management Review, 20*(1), 65–91.

Dorfleitner, G., Utz, S., & Wimmer, M. (2017). Patience pays off – Corporate social responsibility and long-term stock returns. *Journal of Sustainable Finance & Investment, 8*(2), 1–26.

Dowling, J., & Pfeffer, J. (1975). Organizational legitimacy: Social values and organizational behavior. *The Pacific Sociological Review, 18*(1), 122–136.

Drobetz, W., Momtaz, P. P., El Ghoul, S., Guedhami, O., Hackmann, J. P., & Momtaz, P. P. (2024). Entrepreneurial finance and sustainability: Do institutional investors impact the ESG performance of SMEs? *Journal of Business Venturing Insights, 22*, e00498.

Du, P., Huang, S., Hong, Y., & Wu, W. (2022). Can FinTech improve corporate environmental, social, and governance performance? – A study based on the dual path of internal financing constraints and external fiscal incentives. *Frontiers in Environmental Science*. Retrieved January 25, 2023, from https://www.frontiersin.org/articles/10.3389/fenvs.2022.1061454/full

Duchek, S., Raetze, S., & Scheuch, I. (2020). The role of diversity in organizational resilience: A theoretical framework. *Business Research, 13*, 387–423.

Duque-Grisales, E., & Aguilera-Caracuel, J. (2021). Environmental, social and governance (ESG) scores and financial performance of multilatinas: Moderating effects of geographic international diversification and financial slack. *Journal of Business Ethics, 168*(2), 315–334.

Dyck, A., Lins, K. V., Roth, L., & Wagner, H. F. (2019). Do institutional investors drive corporate social responsibility? International evidence. *Journal of Financial Economics, 131*(3), 693–714.

Dzieliński, M., Eugster, F., Sjöström, E., & Wagner, A. F. (2024). *Do firms walk the climate talk?* (Swiss Finance Institute Research Paper, pp. 22–14). Retrieved June 22, 2025, from https://papers.ssrn.com/sol3/papers.cfm?abstract_id=4021061

EBA. (2021). *EBA report on management and supervision of ESG risks for credit institutions and investment firms* (EBA/REP/2021/18). Retrieved January 21, 2022, from https://www.eba.europa.eu/sites/default/documents/files/document_library/Publications/Reports/2021/1015656/EBA%20Report%20on%20ESG%20risks%20management%20and%20supervision.pdf

EBA. (2025). *Final report. Guidelines on the management of environmental, social and governance (ESG) risks*. Retrieved January 10, 2025, from https://www.eba.europa.eu/publications-and media/press-releases/eba-publishes-its-final-guidelines-management-esg-risks

ECB. (2020). *Guide on climate-related and environmental risks: Supervisory expectations relating to risk management and disclosure*. Retrieved January 20, 2021, from https://www.bankingsupervision.europa.eu/ecb/pub/pdf/ssm.202011finalguideonclimate-relatedandenvironmentalrisks~58213f6564.en.pdf

ECB. (2022). *Good practices for climate related and environmental risk management. Observations from the 2022 thematic review*. Retrieved January 8, 2023, from https://www.bankingsupervision.europa.eu/ecb/pub/pdf/ssm.thematicreviewcercompendiumgoodpractices112022~b474fb8ed0.en.pdf

Eccles, R. G. Rajgopal, S., & Xie, J. (2022). *Does ESG Negative Screening Work?* Retrieved Jul 12, 2024, from http://dx.doi.org/10.2139/ssrn.4150524

Eccles, R. G., & Saltzman, D. (2011). Achieving sustainability through integrated reporting. *Stanford Social Innovation Review, 9*(3), 56–61.

Eccles, R. G., Ioannou, I., & Serafeim, G. (2014). The impact of corporate sustainability on organizational processes and performance. *Management Science, 60*(11), 2835–2857.

Eccles, R. G., Krzus, M. P., & Ribot, S. (2015). Models of best practice in integrated reporting 2015. *Journal of Applied Corporate Finance, 27*(2), 103–115.

El Ghoul, S., Guedhami, O., Kwok, C. C., & Mishra, D. R. (2011). Does corporate social responsibility affect the cost of capital? *Journal of Banking & Finance, 35*(9), 2388–2406.

Ellili, N. O. D. (2020). Environmental, social, and governance disclosure, ownership structure and cost of capital: Evidence from the UAE. *Sustainability, 12*(18), 2–23.

Ernst, D., & Woithe, F. (2024). Impact of the environmental, social, and governance rating on the cost of capital: Evidence from the S&P 500. *Journal of Risk and Financial Management, 17*(3), 91.

Fatemi, A., & Fooladi, I. (2020). A primer on sustainable value creation. *Review of Financial Economics, 38*(3), 452–473.

Fatemi, A., Glaum, M., & Kaiser, S. (2018). ESG performance and firm value: The moderating role of disclosure. *Global Finance Journal, 38*, 45–64.

Ferrero-Ferrero, I., Fernández-Izquierdo, M. Á., & Muñoz-Torres, M. J. (2016). The effect of environmental, social and governance consistency on economic results. *Sustainability, 8*(10), 1005.

Fiorillo, P., Meles, A., Ricciardi, A., & Verdoliva, V. (2025). ESG performance and the cost of debt. Evidence from the corporate bond market. *International Review of Financial Analysis, 102*, 104097.

Firmansyah, E. A., Umar, U. H., & Jibril, R. S. (2023). Investigating the effect of ESG disclosure on firm performance: The case of Saudi Arabian listed firms. *Cogent Economics & Finance, 11*(2), 2287923.

Flammer, C. (2021). Corporate green bonds. *Journal of Financial Economics, 142*(2), 499–516.

Flammer, C., Hong, B., & Minor, D. (2019). Corporate governance and the rise of integrating corporate social responsibility criteria in executive compensation: Effectiveness and implications for firm outcomes. *Strategic Management Journal, 40*(7), 1097–1122.

Freeman, R. E. (1994). The politics of stakeholder theory: Some future directions. *Business Ethics Quarterly, 4*(4), 409–421.

Freeman, R. E. (2010). *Strategic management: A stakeholder approach*. Cambridge University Press.

Friede, G., Busch, T., & Bassen, A. (2015). ESG and financial performance: Aggregated evidence from more than 2000 empirical studies. *Journal of Sustainable Finance & Investment, 5*(4), 210–233.

Friedman, M. (2007). The social responsibility of business is to increase its profits. In W. C. Zimmerli, M. Holzinger, & K. Richter (Eds.), *Corporate ethics and corporate governance* (pp. 173–178). Springer.

Friedman, M., Mackey, J., & Rodgers, T. J. (2005). Rethinking the social responsibility of business. A reason debate featuring Milton Friedman, Whole Foods' John Mackey, and Cypress Semiconductor's T.J. Rodgers. *Reason, 37*, 28–37.

Fu, T., & Li, J. (2023). An empirical analysis of the impact of ESG on financial performance: The moderating role of digital transformation. *Frontiers in Environmental Science, 11*, 1256052.

Gallucci, C. (2023, February 10). Mercati finanziari e imprese: il dibattito scientifico in corso nella prospettiva della sostenibilità. *Convegno invernale ADEIMF 2023*.

Gao, C., & Zhang, S. (2025). ESG performance and corporate financialization: A dual perspective of risk management and value creation. *Finance Research Letters, 71*, 106442.

Gao, S., Meng, F., Wang, W., & Chen, W. (2023). Does ESG always improve corporate performance? Evidence from firm life cycle perspective. *Frontiers in Environmental Science, 11*, 1105077.

Gardberg, N. A., & Fombrun, C. J. (2006). Corporate citizenship: Creating intangible assets across institutional environments. *Academy of Management Review, 31*(2), 329–346.

Gherghina, Ş. C. (2024). Corporate finance and Environmental, Social, and Governance (ESG) practices. *Journal of Risk and Financial Management, 17*(7), 308.

Gianfrate, G., Rubin, M., Ruzzi, D., & van Dijk, M. (2024). On the resilience of ESG firms during the COVID-19 crisis: Evidence across countries and asset classes. *Journal of International Business Study, 55*, 1069–1084.

Gibson, K. (2000). The moral basis of stakeholder theory. *Journal of Business Ethics, 26*, 245–257.

Giese, G., Lee, L. E., Melas, D., Nagy, Z., & Nishikawa, L. (2019). Foundations of ESG investing: How ESG affects equity valuation, risk, and performance. *The Journal of Portfolio Management, 45*(5), 69–83.

Gillan, S. L., Koch, A., & Starks, L. T. (2021). Firms and social responsibility: A review of ESG and CSR research in corporate finance. *Journal of Corporate Finance, 66*, 101889.

Ginglinger, E., & Moreau, Q. (2023). Climate risk and capital structure. *Management Science, 69*(12), 7492–7516.

Gjergji, R., Vena, L., Sciascia, S., & Cortesi, A. (2021). The effects of environmental, social and governance disclosure on the cost of capital in small and medium enterprises: The role of family business status. *Business Strategy and the Environment, 30*(1), 683–693.

Godfrey, P. C. (2005). The relationship between corporate philanthropy and shareholder wealth: A risk management perspective. *Academy of Management Review, 30*(4), 777–798.

Gonçalves, T. C., Dias, J., & Barros, V. (2022). Sustainability performance and the cost of capital. *International Journal of Financial Studies, 10*(3), 63.

Goss, A., & Roberts, G. S. (2011). The impact of corporate social responsibility on the cost of bank loans. *Journal of Banking & Finance, 35*(7), 1794–1810.

Gupta, K. (2018). Environmental sustainability and implied cost of equity: International evidence. *Journal of Business Ethics, 147*, 343–365.

Han, W., & Wu, D. (2024). ESG ratings, business credit acquisition, and corporate value. *International Review of Financial Analysis, 95*(Part A), 103376.

Hasan, I., Hoi, C. K., Wu, Q., & Zhang, H. (2017). Social capital and debt contracting: Evidence from bank loans and public bonds. *Journal of Financial and Quantitative Analysis, 52*(3), 1017–1047.

Hauptmann, C. (2017). *Corporate sustainability performance and bank loan pricing: It pays to be good, but only when banks are too* (SSRN Working Paper). Retrieved August 7, 2023, from https://papers.ssrn.com/sol3/papers.cfm?abstract_id=3067422

Ho, L., Bai, M., Lu, Y., & Qin, Y. (2021). The effect of corporate sustainability performance on leverage adjustments. *The British Accounting Review, 53*(5), 100989.

Hoepner, A. G. F., Oikonomou, I., Sautner, Z., Laura, T., Starks, L. T., & Zhou, X. Y. (2024). ESG shareholder engagement and downside risk. *Review of Finance, 28*(2), 483–510.

Hornuf, L., & Yüksel, G. (2024). The performance of socially responsible investments: A meta-analysis. *European Financial Management, 30*(2), 1012–1061.

Houston, J. F., & Shan, H. (2022). Corporate ESG Profiles and Banking Relationships. *The Review of Financial Studies, 35*(7), 3373–3417.

Hu, S., Chen, P., & Zhang, C. (2025). How does green finance reform affect corporate ESG greenwashing behavior? *International Review of Financial Analysis, 102*, 104037.

Huang, D. Z. X. (2019). Environmental, social and governance (ESG) activity and firm performance: A review and consolidation. *Accounting & Finance, 61*(1), 335–360.

Huang, Z., Shi, Y., & Jia, M. (2024). Greenwashing: A systematic literature review. *Accounting and Finance, 65*(1), 857–882.

Hughes, A., Urban, M. A., & Wójcik, D. (2021). Alternative ESG ratings: How technological innovation is reshaping sustainable investment. *Sustainability, 13*(6), 1–23.

ICMA. (2024). *Sustainability-linked bond principles. Voluntary process guidelines.* Retrieved December 28, 2024, from https://www.icmagroup.org/assets/documents/Sustainable-finance/2024-updates/Sustainability-Linked-Bond-Principles-June-2024.pdf

Ioannou, I., & Serafeim, G. (2017). *The consequences of mandatory corporate sustainability reporting* (Harvard Business School Research Working Paper No. 11-100). Retrieved November 15, 2024, from https://ssrn.com/abstract=1799589

Jensen, M. C. (2002). Value maximisation, stakeholder theory, and the corporate objective function. *European Financial Management, 7*(3), 297–317.

Jensen, J. C., & Berg, N. (2012). Determinants of traditional sustainability reporting versus integrated reporting. An institutionalist approach. *Business Strategy and the Environment, 21*(5), 299–316.

Jensen, M. C., & Meckling, W. H. (1976). Theory of the firm: Managerial behavior, agency costs and ownership structure. *Journal of Financial Economics, 3*(4), 305–360.

Jo, H., & Harjoto, M. A. (2011). Corporate governance and firm value: The impact of corporate social responsibility. *Journal of Business Ethics, 103*(3), 351–383.

Johnson, R. (2020). The link between environmental, social and corporate governance disclosure and the cost of capital in South Africa. *Journal of Economic and Financial Sciences, 13*(1), 1–12.

Karolyi, G. A., Kim, D., & Liao, R. (2020). The theory and practice of investor relations: A global perspective. *Management Science, 66*(10), 4746–4771.

Katelouzou, D., & Klettner, A. (2022). Sustainable finance and stewardship: Unlocking stewardship's sustainability potential. In D. Katelouzou & D. W. Puchniak (Eds.), *Global shareholder stewardship: Complexities, challenges, and possibilities* (pp. 549–571). Cambridge University Press.

Khan, M., Serafeim, G., & Yoon, A. (2016). Corporate sustainability: First evidence on materiality. *Accounting Review, 91*(6), 1697–1724.

Khan, M., Serafeim, G., & Yoon, A. (2021). *ESG and financial performance: Uncovering the relationship by aggregating evidence from 1,000 plus studies published between 2015–2020.* NYU Stern Center for Sustainable Business and Rockefeller Asset Management. Retrieved January 15, 2023, from https://www.stern.nyu.edu/sites/default/files/assets/documents/NYU-RAM_ESG-Paper_2021%20Rev_0.pdf

Kostova, T., & Zaheer, S. (1999). Organizational legitimacy under conditions of complexity: The case of the multinational enterprise. *Academy of Management Review, 24*(1), 64–81.

Kotsantonis, S., & Serafeim, G. (2019). Four things no one will tell you about ESG data. *Journal of Applied Corporate Finance, 31*(2), 50–58.

KPMG. (2023). *ESG due diligence. A strategic imperative in mergers & acquisitions*. Retrieved January 25, 2024, from https://assets.kpmg.com/content/dam/kpmg/qa/pdf/2023/09/esg-due-diligence.pdf

Kraus, A., & Litzenberger, R. H. (1973). A state-preference model of optimal financial leverage. *The Journal of Finance, 28*(4), 911–922.

Krueger, P., Sautner, Z., & Starks, L. T. (2020). The importance of climate risks for institutional investors. *The Review of Financial Studies, 33*(3), 1067–1111.

Krüger, P. (2015). Corporate goodness and shareholder wealth. *Journal of Financial Economics, 115*(2), 304–329.

Larcker, D. F., Tayan, B., & Watts, E. M. (2022). Seven myths of ESG. *European Financial Management, 28*(4), 869–882.

Lee Brown, D., Guidry, R. P., & Patten, D. M. (2009). Sustainability reporting and perceptions of corporate reputation: An analysis using fortune. *Sustainability, Environmental Performance and Disclosures, 4*, 83–104.

Lee, S.-P., & Isa, M. (2022). Environmental, Social and Governance (ESG) practices and financial performance of Shariah-compliant companies in Malaysia. *Journal of Islamic Accounting and Business Research, 14*(2), 295–314.

Lengnick-Hall, C. A., & Beck, T. E. (2005). Adaptive fit versus robust transformation: How organizations respond to environmental change. *Journal of Management, 31*(5), 738–757.

Leuz, C., & Wysocki, P. D. (2016). The economics of disclosure and financial reporting regulation: Evidence and suggestions for future research. *Journal of Accounting Research, 54*(2), 525–622.

Li, Y., & Zhang, Z. (2023). Corporate climate risk exposure and capital structure: Evidence from Chinese listed companies. *Finance Research Letters, 51*, 103488.

Li, Y., Gong, M., Zhang, X. Y., & Koh, L. (2018). The impact of environmental, social, and governance disclosure on firm value: The role of CEO power. *The British Accounting Review, 50*(1), 60–75.

Li, C., He, S., Tian, Y., Sun, S., & Ning, L. (2022). Does the bank's FinTech innovation reduce its risk-taking? Evidence from China's banking industry. *Journal of Innovation & Knowledge, 7*(3), 100219.

Li, Y., Zhao, X., & Yang, X. (2025). ESG performance and corporate environmental investment: Incentive or inhibition? *Finance Research Letters, 75*, 106814.

Liang, H., & Renneboog, L. (2020). *Corporate social responsibility and sustainable finance: A review of the literature* (European Corporate Governance Institute – Finance Working Paper No. 701/2020). Retrieved September 18, 2024, from https://papers.ssrn.com/sol3/papers.cfm?abstract_id=3698631

Liaw, T. (2020). Survey of green bond pricing and investment performance. *Journal of Risk and Financial Management, 13*(9).

Lin, H., Gu, X., & Bao, X. (2025). Corporate ESG performance and bankruptcy risk. *Finance Research Letters, 76*, 106987.

Lins, K. V., Servaes, H., & Tamayo, A. (2017). Social capital, trust, and firm performance: The value of corporate social responsibility during the financial crisis. *The Journal of Finance, 72*(4), 1785–1824.

Lintner, J. (1965). The valuation of risk assets and the selection of risky investments in stock portfolios and capital budgets. *The Review of Economics and Statistics, 47*(1), 13–37.

Liu, Y.-Y., & Lee, P.-S. (2025). The effect of Environmental, Social, and Governance (ESG) on the persistence of firm value: Evidence from survival analysis. *Accounting and Auditing, 1*(1), 4.

Long, H., Chiah, M., Cakici, N., Zaremba, A., & Bilgin, M. H. (2024). ESG investing in good and bad times: An international study. *Journal of International Financial Markets, Institutions and Money, 91*, 101916.

Lopez-de-Silanes, F., McCahery, J. A., & Pudschedl, P. C. (2020). ESG performance and disclosure: A cross-country analysis. *Singapore Journal of Legal Studies*, 217–241.

Lopez-de-Silanes, F., McCahery, J. A., & Pudschedl, P. C. (2024). Institutional investors and ESG preferences. *Corporate Governance: An International Review, 32*(6), 1060–1086.

Mackey, A., Mackey, T. B., & Barney, J. B. (2007). Corporate social responsibility and firm performance: Investor preferences and corporate strategies. *The Academy of Management Review, 32*(3), 817–835.

Mahanta, A., Sahu, N. C., Behera, P. K., & Kumar, P. (2024). Variations in financial performance of firms with ESG integration in business: The mediating role of corporate efficiency using DEA. *Green Finance, 6*(3), 518–562.

Martielli, F., Battisti, E., Gonzalez-Cruz, T. F., & Salvi, A. (2025). Carbon credits and financial performance: Exploring the moderating role of CSR strategy and corporate governance practices. *Research in International Business and Finance, 77*(Part B), 102919.

Martin-Melero, I., Gomez-Martinez, R., Medrano-Garcia, M. L., & Hernandez-Perlines, F. (2025). Comparison of sectorial and financial data for ESG scoring of mutual funds with machine learning. *Financial Innovation, 11*(84).

Matthiesen, M. L., & Salzmann, A. J. (2017). Corporate social responsibility and firms' cost of equity: How does culture matter? *Cross Cultural & Strategic Management, 24*(1), 105–124.

Mcwilliams, A., & Siegel, D. (1997). The role of money managers in assessing corporate social Reponsibility research. *The Journal of Investing, 6*(4), 98–107.

Menicucci, E., & Paolucci, G. (2023). ESG dimensions and bank performance: An empirical investigation in Italy. *Corporate Governance, 23*(3), 563–586.

Miller, D., & Friesen, P. H. (1984). A longitudinal study of the corporate life cycle. *Management Science, 30*(10), 1161–1183.

Miralles-Quirós, M. M., Jerez-Barroso, L., Miralles-Quirós, J. L., & Redondo-Cordero, M. (2023). Extremadura's small and medium livestock enterprises and sustainable development goals. In W. Leal Filho, M. A. P. Dinis, S. Moggi, E. Price, & A. Hope (Eds.), *SDGs in the European region. Implementing the UN sustainable development goals – Regional perspectives*. Springer.

Mneimneh, F., Al Kodsi, M., Chamoun, M., & Basharoush, M. (2023). How can green energy technology innovations improve the carbon-related environmental dimension of ESG rating? *Circular Economy and Sustainability, 3*, 2183–2199.

Morgan Stanley. (2015). *Sustainable reality: Understanding the performance of sustainable investment strategies*. Retrieved February 21, 2022, from https://www.riacanada.ca/research/sustainable-reality-understanding-the-performance-of-sustainable-investment-strategies/

Mossin, J. (1966). Equilibrium in a capital asset market. *Econometrica, 34*(4), 768–783.

Muchenje, L. T., Coupe, T., & Dang, H. D. (2025). Credit development and ESG performance: Cross-country evidence. *Research in International Business and Finance, 78*, 103007.

Murata, R., & Hamori, S. (2021). ESG disclosures and stock price crash risk. *Journal of Risk and Financial Management, 14*(70), 1–20.

Mutarindwa, S., Schäfer, D., & Stephan, A. (2024). Certification against greenwashing in nascent bond markets: Lessons from African ESG bonds. *Eurasian Economic Review, 14*, 149–173.

Myers, S. C., & Majluf, N. S. (1984). Corporate financing and investment decisions when firms have information that investors do not have. *Journal of Financial Economics, 13*(2), 187–221.

Naeem, M., Ullah, H., & Jan, S. (2021). The impact of ESG practices on firm performance: Evidence from emerging countries. *Indian Journal of Economics and Business, 20*(1), 731–750.

Naeem, M. A., Yousaf, I., Karim, S., Tiwari, A. K., & Farid, S. (2023). Comparing asymmetric price efficiency in regional ESG markets before and during COVID-19. *Economic Modelling, 118*, 106095.

Narula, R., Rao, P., & Rao, A. A. (2023). Impact of ESG on firm value: A conceptual review of the literature. *Journal of Social and Economic Development, 25*, 162–179.

Naseer, M. M., Khan, M. A., Bagh, T., Guo, Y., & Zhu, X. (2024). Firm climate change risk and financial flexibility: Drivers of ESG performance and firm value. *Borsa Istanbul Review, 24*(1), 106–117.

Nazir, M., Akbar, M., Akbar, A., Poulovo, P., Hussain, A., & Qureshi, M. A. (2022). The nexus between corporate environment, social, and governance performance and cost of capital: Evidence from top global tech leaders. *Environmental Science and Pollution Research, 29*, 22623–22636.

Nekhili, M., Boukadhaba, A., & Nagati, H. (2021). The ESG–financial performance relationship: Does the type of employee board representation matter? *Corporate Governance: An International Review, 29*(2), 134–161.

Neukirchen, D., Engelhardt, N., Krause, M., & Posch, P. N. (2023). The value of (private) investor relations during the COVID-19 crisis. *Journal of Banking & Finance, 147*, 1–26.

Newton, D. P., Ongena, S., Xie, R., & Zhao, B. (2024). Firm ESG reputation risk and debt choice. *European Financial Management, 30*(4), 2071–2094.

Nguyen, P. (2020). Board gender diversity and cost of equity. *Applied Economics Letters, 27*(18), 1522–1526.

Nguyen, J. H., & Phan, H. V. (2020). Carbon risk and corporate capital structure. *Journal of Corporate Finance, 64*, 101713.

Nian, H., & Said, F. F. (2024). The impact of ESG on firm risk and financial performance: A systematic literature review. *Journal of Scientometric Research, 13*(3s), s144–s155.

Nofsinger, J., & Varma, A. (2014). Socially responsible funds and market crises. *Journal of Banking & Finance, 48*, 180–193.

NORDSIP. (2019). *Sustainable index funds & ETFS*. Handbook series. Retrieved May 24, 2024, from https://nordsip.com/wp-content/uploads/2019/04/Handbook-Sustainable-Index-Funds-ETFs-spread-1.pdf

Nurlanovich, B. J. (2025). Social responsibility and ESG in SMEs for entrepreneurship growth: Mapping study. *BIO Web of Conferences, 175*, 06003. Retrieved May 30, 2025, from https://www.bio-conferences.org/articles/bioconf/pdf/2025/26/bioconf_istakcos2024_06003.pdf

OECD. (2022). *Financing SMEs for sustainability: Drivers, constraints and policies* (OECD SME and Entrepreneurship Papers, 35). Retrieved May 30, 2025, from https://www.oecd-ilibrary.org/economics/financing-smes-for-sustainability_a5e94d92-en

Palacin-Bossa, G. D., Alvear-Montoya, L. G., & Macías-Jiménez, M. A. (2024). Exploring the relationship between ESG, financial performance, and corporate reputation using ANOVA: The case of Colombian companies. *Procedia Computer Science, 241*, 552–557.

Paolone, F., & Pozzoli, M. (2021). La valutazione d'azienda verso una nuova direzione dopo anni di "allineamento generalizzato". Il contributo dell'ESG score. *La Valutazione delle Aziende, 4*, 2–9.

Perez-Batres, L. A., Doh, J. P., Miller, V. V., & Pisani, M. J. (2012). Stakeholder pressures as determinants of CSR strategic choice: Why do firms choose symbolic versus substantive self-regulatory codes of conduct? *Journal of Business Ethics, 110*, 157–172.

Perrini, F., Russo, A., Tencati, A., & Vurro, C. (2011). Deconstructing the relationship between corporate social and financial performance. *Journal of Business Ethics, 102*, 59–76.

Pfeffer, J., & Salancik, G. R. (1978). *The external control of organizations: A resource dependence perspective*. Harper & Row.

Pham, H. S. T., & Tran, H. T. (2020). CSR disclosure and firm performance: The mediating role of corporate reputation and moderating role of CEO integrity. *Journal of Business Research, 120*, 127–136.

Pizzi, S., Corbo, L., & Caputo, A. (2021). Fintech and SMEs sustainable business models: Reflections and considerations for a circular economy. *Journal of Cleaner Production, 281*, 125217.

Postiglione, M., Carini, C., & Falini, A. (2024). ESG and firm value: A hybrid literature review on cost of capital implications from Scopus database. *Corporate Social Responsibility and Environmental Management, 31*(6), 6457–6480.

Priem, R., & Gabellone, A. (2024). The impact of a firm's ESG score on its cost of capital: Can a high ESG score serve as a substitute for a weaker legal environment. *Sustainability Accounting, Management and Policy Journal, 15*(3), 676–703.

Quintiliani, A. (2008). *Appunti di finanza aziendale*. Aracne.

Quintiliani, A. (2016). *Internal rating systems e soft information. Il ruolo degli intangibili e del contesto territoriale nella valutazione del merito creditizio delle PMI*. Franco Angeli.

Quintiliani, A. (2021). *L'impresa ai tempi del Covid-19. Pianificare è meglio che sanificare*. G. Giappichelli Editore.

Quintiliani, A. (2022). ESG and firm value. *Accounting and Finance Research, 11*(4), 37–47.

Quintiliani, A. (2024). Why and how ESG practices shape firms' financial performance. *International Journal of Academic Research in Accounting, Finance and Management Sciences, 14*(4), 958–987.

Quintiliani, A. (2025). Small and medium-sized enterprises and sustainable transition: Role of FinTech in a country's banking ecosystem. *Research in International Business and Finance, 73*(Part A), 102625, 1–15.

Rees, W., & Rodionova, T. (2015). The influence of family ownership on corporate social responsibility: An international analysis of publicly listed companies. *Corporate Governance: An International Review, 23*(3), 184–202.

Renneboog, L., Horst, J. T., & Zhang, C. (2008). Socially responsible investments: Institutional aspects, performance, and investor behavior. *Journal of Banking & Finance, 32*(9), 1723–1742.

Reshetnikova, I., & Sanak-Kosmowska, K. (2023). Corporate social responsibility of companies in the context of Russian military aggression in Ukraine. *Marketing and Management of Innovations, 2*, 138–151.

Rhou, Y., Singal, M., & Koh, Y. (2016). CSR and financial performance: The role of CSR awareness in the restaurant industry. *International Journal of Hospitality Management, 57*, 30–39.

Ricci, O., Santilli, G., Scardozzi, G., & Lopes, F. S. S. (2024). ESG resilience in conflictual times. *Research in International Business and Finance, 71*, 102411.

Rong, X., & Kim, M.-I. (2024). ESG and the cost of debt: Role of media coverage. *Sustainability, 16*(12), 4993.

Roy, J. K., & Vasa, L. (2025). Financial technology and environmental, social and governance in sustainable finance: A bibliometric and thematic content analysis. *Discover Sustainability, 6*(1), 148.

Sahin, Ö., Bax, K., Czado, C., & Paterlini, S. (2022). Environmental, social, governance scores and the missing pillar – Why does missing information matter? *Corporate Social Responsibility and Environmental Management, 29*(5), 1782–1798.

Saputra, A., & Rahman, A. (2024). ESG performance and its impact on mitigating cost of capital: Evidence from Southeast Asia. *E-Jurnal Akuntansi, 34*(8), 2054–2072.

Sautner, Z., Van Lent, L., Vilkov, G., & Zhang, R. (2023). Pricing climate change exposure. *Management Science, 69*(12), 7540–7561.

Schenker, J. (2020). *The future of finance after COVID: Technology and trends disrupting the post-pandemic financial world.* Prestige Professional Publishing.

Scott, W. R. (2001). *Institutions and organizations* (2nd ed.). Sage Publications.

Seth, R., & Mahenthiran, S. (2022). Impact of dividend payouts and corporate social responsibility on firm value – Evidence from India. *Journal of Business Research, 146*, 571–581.

Sharfman, M. P., & Fernando, C. S. (2008). Environmental risk management and the cost of capital. *Strategic Management Journal, 29*(6), 569–592.

Sharpe, W. F. (1964). Capital asset prices: A theory of market equilibrium under conditions of risk. *The Journal of Finance, 19*(3), 425–442.

Shi, Y., Zheng, S., Xiao, P., Zhen, H., & Wu, T. (2024). ESG performance and cost of debt. *China Journal of Accounting Research, 17*, 100390.

Simnett, R., Vanstraelen, A., & Chua, W. F. (2009). Assurance on sustainability reports: An international comparison. *The Accounting Review, 84*(3), 937–967.

Singh, D. A., & Gaur, A. S. (2009). Business group affiliation, firm governance, and firm performance: Evidence from China and India. *Corporate Governance. An International Review, 17*(4), 411–425.

Singh, P. J., Sethuraman, K., & Lam, J. Y. (2017). Impact of corporate social responsibility dimensions on firm value: Some evidence from Hong Kong and China. *Sustainability, 9*(9), 1–24.

Sneideriene, A., & Legenzova, R. (2025). Greenwashing prevention in environmental, social, and governance (ESG) disclosures: A bibliometric analysis. *Research in International Business and Finance, 74*, 102720.

Spence, M. (1973). Job market signaling. *The Quarterly Journal of Economics, 87*(3), 355–374.

Steyn, M. (2014). Organisational benefits and implementation challenges of mandatory integrated reporting: Perspectives of senior executives at South African listed companies. *Sustainability Accounting, Management and Policy Journal, 5*(4), 476–503.

Stiglitz, J. E., & Weiss, A. (1981). Credit rationing in markets with imperfect information. *The American Economic Review, 71*(3), 393–410.

Suchman, M. C. (1995). Managing legitimacy: Strategic and institutional approaches. *The Academy of Management Review, 20*(3), 571–610.

Suto, M., & Takehara, H. (2017). CSR and cost of capital: Evidence from Japan. *Social Responsibility Journal, 13*(4), 798–816.

Tamasiga, P., Onyeaka, H., Bakwena, M., & Ouassou, E. (2024). Beyond compliance: Evaluating the role of environmental, social and governance disclosures in enhancing firm value and performance. *SN Business & Economics, 4*(118).

Tamimi, N., & Sebastianelli, R. (2017). Transparency among S&P 500 companies: An analysis of ESG disclosure scores. *Management Decision, 55*(8), 1660–1680.

Tang, D. Y., & Zhang, Y. (2020). Do shareholders benefit from green bonds? *Journal of Corporate Finance, 61*, 101427.

Tawfiq, T. T., Tawaha, H., Tahtamouni, A., & Almasria, N. A. (2024). The influence of environmental, social, and governance disclosure on capital structure: An investigation of leverage and WACC. *Journal of Risk and Financial Management, 17*(12), 570.

Teece, D. J. (2007). Explicating dynamic capabilities: The nature and microfoundations of (sustainable) enterprise performance. *Strategic Management Journal, 28*(13), 1319–1350.

Thuy, H. X., Khuong, N. V., Anh, L. H. T., & Quyen, P. N. (2024). Effect of corporate governance on corporate social responsibility in Vietnam: State-ownership as the moderating role. *Journal of Financial Reporting and Accounting, 22*(3), 701–727.

Treynor, J. L. (1961). *Market Value, Time, and Risk.* Retrieved May 18, 2024, from http://dx.doi.org/10.2139/ssrn.2600356.

van Duuren, E., Plantinga, A., & Scholtens, B. (2016). ESG integration and the investment management process: Fundamental investing reinvented. *Journal of Business Ethic, 138*, 525–533.

Velte, P. (2017). Does ESG performance have an impact on financial performance? Evidence from Germany. *Journal of Global Responsibility, 8*(2), 169–178.

Verheyden, T., Eccles, R. G., & Feiner, A. (2016). ESG for all? The impact of ESG screening on return, risk, and diversification. *Journal of Applied Corporate Finance, 28*(2), 47–55.

Waddock, S. A., & Graves, S. B. (1997). The corporate social performance-financial performance link. *Strategic Management Journal, 18*(4), 303–319.

Wahyuni, R., Febriyanti, B., Laila, G., Sunaryo, D., & Adiyanto, Y. (2024). Sustainability based financial risk management strategies for Long term resilience: A systematic review. *Indo-Fintech Intellectuals: Journal of Economics and Business, 4*(5), 2625–2639.

Walker, T., Fernandes, A., & Karami, M. (2024). Corporate resilience during crises and the role of ESG factors: Evidence from the COVID-19 pandemic. *Finance Research Letters, 69*(PB), 106168.

Wamba, S. F., & Queiroz, M. M. (2022). A framework based on blockchain, artificial intelligence, and big data analytics to leverage supply chain resilience considering the COVİD-19. *Ifac-Papersonline, 55*(10), 2396–2401.

Wang, D., Peng, K., Tang, K., & Wu, Y. (2022). Does Fintech development enhance corporate ESG performance? Evidence from an emerging market. *Sustainability, 14*(24), 1–21.

Wang, K., Yu, S., Mei, M., Yang, X., Peng, G., & Lv., B. (2023). ESG performance and corporate resilience: An empirical analysis based on the capital allocation efficiency perspective. *Sustainability, 15*(23), 16145.

Wang, N., Pan, H., Feng, Y., & Du, S. (2024). How do ESG practices create value for businesses? Research review and prospects. *Sustainability Accounting, Management and Policy Journal, 15*(5), 1155–1177.

Wei, D., & Zeng, Y. (2025). Enhancing risk management in inclusive ESG investment portfolios in financial markets through big data analysis. *Journal of Computational Methods in Sciences and Engineering, 25*(3), 2275–2287.

Welsh, H. (2018). *Social, environmental & sustainable governance shareholder proposals in 2018* (SII Working Paper). Retrieved January 11, 2025, from http://www.sec.gov/comments/4-725/4725-4636528-176443.pdf

Wong, W. C., Batten, J. A., Ahmad, A. H., Mohamed-Arshad, S. B., Nordin, S., & Adzis, A. A. (2021). Does ESG certification add firm value? *Finance Research Letters, 39*, 101593.

Wu, K., Zhang, Y., Chen, Y., & Li, M. (2025). Impact of digital finance on corporate ESG. *International Review of Financial Analysis, 104*(Part A), 104259.

Yadav, A., & Asongu, S. A. (2025). The role of ESG performance in moderating the impact of financial distress on company value: Evidence of wavelet-enhanced quantile regression with Indian companies. *Business Strategy and the Environment, 34*(3), 2782–2798.

Yang, C., Zhu, C., & Albitar, K. (2024). ESG ratings and green innovation: A U-shaped journey towards sustainable development. *Business Strategy and the Environment, 33*(5), 4108–4129.

Yavuz, M. S., Tatlı, H. S., Bozkurt, G., & Öngel, G. (2025). Does ESG performance have an impact on financial performance? Evidence from Turkey. *Journal of Entrepreneurship, Management and Innovation, 21*(1), 24–42.

Yoon, B., Lee, J., & Byun, R. (2018). Does ESG performance enhance firm value? Evidence from Korea. *Sustainability, 10*(10), 1–18.

Yu, E. P., Guo, C. Q., & Luu, B. V. (2018). Environmental, social and governance transparency and firm value. *Business Strategy and the Environment, 27*(7), 987–1004.

Zhou, Z., & Wu, K. (2023). Does climate risk exposure affect corporate leverage adjustment speed? International evidence. *Journal of Cleaner Production, 389*, 136036.

Zioło, M., Bąk, I., & Spoz, A. (2023). Theoretical framework of sustainable value creation by companies. What do we know so far? *Corporate Social Responsibility and Environmental Management, 30*(5), 2344–2361.

MIX
Papier aus verantwortungsvollen Quellen
Paper from responsible sources
FSC® C105338

If you have any concerns about our products,
you can contact us on
ProductSafety@springernature.com

In case Publisher is established outside the EU,
the EU authorized representative is:
Springer Nature Customer Service Center GmbH
Europaplatz 3, 69115 Heidelberg, Germany

Printed by Libri Plureos GmbH
in Hamburg, Germany